T0337886

Nanoimprint Technology

Microsystem and Nanotechnology Series

Series Editors: *Ron Pethig and Horacio Dante Espinosa*

Nanoimprint Technology: Nanotransfer for Thermoplastic and Photocurable Polymers
Taniguchi, Ito, Mizuno and Saito, August 2013

Nano and Cell Mechanics: Fundamentals and Frontiers
Espinosa and Bao, January 2013

Digital Holography for MEMS and Microsystem Metrology
Asundi, July 2011

Multiscale Analysis of Deformation and Failure of Materials
Fan, December 2010

Fluid Properties at Nano/Meso Scale
Dyson et al., September 2008

Introduction to Microsystem Technology
Gerlach, March 2008

AC Electrokinetics: Colloids and Nanoparticles
Morgan and Green, January 2003

Microfluidic Technology and Applications
Koch et al., November 2000

Nanoimprint Technology
Nanotransfer for Thermoplastic and Photocurable Polymers

Editors

Jun Taniguchi

Tokyo University of Science, Japan

Hiroshi Ito

Yamagata University, Japan

Jun Mizuno

Waseda University, Japan

Takushi Saito

Tokyo Institute of Technology, Japan

This edition first published 2013

Registered office
John Wiley & Sons Ltd, The Atrium, Southern Gate, Chichester, West Sussex, PO19 8SQ, United Kingdom

For details of our global editorial offices, for customer services and for information about how to apply for permission to reuse the copyright material in this book please see our website at www.wiley.com.

Library of Congress Cataloging-in-Publication Data

Nanoimprint technology : nanotransfer for thermoplastic and photocurable polymer / edited by Jun Taniguchi, Hiroshi Ito, Jun Mizuno, Takushi Saito.
pages cm
Includes bibliographical references and index.
ISBN 978-1-118-35983-9 (cloth)
1. Nanoimprint lithography. 2. Nanolithography–Materials. 3. Plastics–Molding. 4. Polymers–Thermal properties. 5. Thermoplastics. 6. Microfluidics. 7. Transfer-printing. I. Taniguchi, Jun.
TK7874.843.N36 2013
621.381–dc23

2013007112

A catalogue record for this book is available from the British Library.

Print ISBN: 978-1-118-35983-9

Typeset in 10/12pt Palatino by Laserwords Private Limited, Chennai, India

Printed and bound in Malaysia by Vivar Printing Sdn Bhd

1 2013

Contents

About the Editors

Jun Taniguchi is an Associate Professor at the Department of Applied Electronics, Tokyo University of Science (Tokyo, Japan). He received BE, ME, and PhD degrees from Tokyo University of Science, in 1994, 1996, and 1999, respectively. From 1999 to 2013, he was with the Department of Applied Electronics, Tokyo University of Science. His research interests include electron beam lithography for nanoimprint molding, nanoimprint lithography, roll-to-roll nanoimprint lithography, and nanotechnology applications such as optical devices and moth-eye structures (junt@te.noda.tus.ac.jp).

Hiroshi Ito graduated from the Department of Polymeric Materials and Engineering at Yamagata University (Yamagata, Japan). He received his Master's degree in engineering from Yamagata University, in 1990. After graduation, he joined Oki Electric Industry Co., Ltd (Tokyo, Japan). In 1993, he became an Assistant Professor at Tsuruoka National College of Technology (Yamagata, Japan), and received his PhD from Yamagata University, in 1996. In this year, he also became an Assistant Professor at the Tokyo Institute of Technology (Tokyo, Japan). In 2007, he became an Associate Professor at Yamagata University. In 2010, he became a Professor at Yamagata University. He is now Chair of the

Department of Organic Device Engineering and the Department of Organic Materials Engineering (PhD program). He is also Director of Research at the Center for Advanced Processing GREEN Materials, Yamagata University (ihiroshi@yz.yamagata-u.ac.jp).

Jun Mizuno received his PhD in applied physics from Tohoku University (Miyagi, Japan) in 2000. He is currently an Associate Professor at Waseda University (Tokyo, Japan) and works at the Nanotechnology Research Center. His current interests are MEMS/NEMS technology, bonding technology at low temperature using plasma activation or excimer laser irradiation, printed electronics, and composite technology for UV/heat nanoimprint lithography combined with electrodeposition (mizuno@waseda.jp).

Takushi Saito is a member of the Department of Mechanical and Control Engineering at the Tokyo Institute of Technology (Tokyo, Japan). He received his PhD in engineering from the Tokyo Institute of Technology in 1996. He began his academic career as a post-doctoral researcher at the University of Minnesota (MN, USA) in 1997. He then became an Assistant Professor at the Tokyo Institute of Technology in 1998. Since June 2002, he has been an Associate Professor at the Tokyo Institute of Technology. His current research topics include visualization and measurement of polymer processing, laser-assisted manufacturing processes, and the development of heat transfer control techniques in material processing (tsaito@mep.titech.ac.jp).

List of Contributors

Hiroshi Goto
Toshiba Machine Co., Ltd, Japan
(goto.hiroshi@toshiba-machine.co.jp)

Hiroshi Ito
The Japan Steel Works, Ltd, Hiroshima Research Laboratory, Japan
(hiroshi_ito@jsw.co.jp)

Mitsunori Kokubo
Toshiba Machine Co., Ltd, Japan
(kokubo.mitunori@toshiba-machine.co.jp)

Kenichi Kotaki
SmicS Co., Ltd, Japan
(kotaki@smics-jp.com)

Hiroto Miyake
Daicel Corporation, Planning R & D Management, Tokyo Head Office, Japan
(hr_miyake@jp.daicel.com)

Masao Otaki
Toppan Printing Co., Ltd, Japan

Nobuji Sakai
Samsung R&D Institute, Japan
(n.sakai@samsung.com)

Hidetoshi Shinohara
Toshiba Machine Co., Ltd, Japan
(shinohara.hidetoshi@toshiba-machine.co.jp)

Gaku Suzuki
Toppan Printing Co., Ltd, Japan
(gaku.suzuki@toppan.co.jp)

Kentaro Tsunozaki
Asahi Glass Co., Ltd, Research Center, Japan
(kentaro-tsunozaki@agc.co.jp)

Noriyuki Unno
Tokyo University of Science, Faculty of Industrial Science and Technology, Department of Applied Electronics, Japan
(n.unno@rs.tus.ac.jp)

Kazutoshi Yakemoto
The Japan Steel Works, Ltd, Hiroshima Research Laboratory, Japan
(kazutoshi_yakemoto@jsw.co.jp)

Norio Yoshino
Tokyo University of Science, Faculty of Engineering, Department of Industrial Chemistry, Japan
(yoshino@ci.kagu.tus.ac.jp)

Series Preface

The Microsystem and Nanotechnology book series provides a thorough contextual summary of the current methods used in micro- and nanotechnology research and how these advances are influencing many scientific fields of study and practical application. Readers of these books are guided to learn the fundamental principles necessary for the topic, while finding many examples that are representative of the application of these fundamental principles. This approach ensures that the books are appropriate for readers with varied backgrounds and useful for self-study or as classroom materials.

Micro- and nanoscale materials, fabrication techniques, and metrology methods are the basis for many modern technologies. Several books in this series, including *Introduction to Microsystem Technology* by Gerlach and Dotzel, *Microfluidic Technology and Applications* edited by Koch, Evans, and Brunnschweiler, and *Fluid Properties at Nano/Meso Scale* by Dyson, Ransing, P. Williams, and R. Williams, provide a resource for building a scientific understanding of the field. Multiscale modeling, an important aspect of microsystem design, is extensively reviewed in *Multiscale Analysis of Deformation and Failure of Materials* by Jinghong Fan. Modern topics in mechanics are covered in *Nano and Cell Mechanics: Fundamentals and Frontiers* edited by Espinosa and Bao. Specific implementations and applications are presented in *AC Electrokinetics: Colloids and Nanoparticles* by Morgan and Green, *Digital Holography for MEMS and Microsystem Metrology* edited by Asundi.

This book, edited by Jun Taniguchi, presents the fundamental methods of nanoimprint technologies and the principles of fabrication and materials selection that are essential for their successful implementation. Included in this work are examples of theoretical modeling of the physical phenomena that govern micro- and nanofabrication and the invaluable insight they provide for informing process design and parameters.

Horacio D. Espinosa
Ron Pethig

Preface

The technique of nanoscale pattern transfer technology using a mold has attracted attention because this technology makes nanotechnology industries and applications possible. This field of technology has evolved rapidly, year by year. However, because of these rapid advances, it is difficult to keep up with the technological trends and the latest cutting-edge methods. In order to fully understand these pioneering technologies, comprehension of the basic science and an overview of the techniques is required. In this book, the latest nanotransfer science – based on polymer behavior and polymer fluid dynamics – is described in detailed but easy-to-understand language. Based on their physical science, injection molding and nanoimprint lithography are explored. These exemplifications of concrete methods will help the reader to create an accurate picture of nanofabrication. Furthermore, the newest cutting-edge nanotransfer technologies and applications are also described. We hope the reader will benefit from knowledge of these new technologies and be left with a basic comprehension of nanotransfer mechanisms and methods.

Jun Taniguchi

1

What is a Nanoimprint?

Jun Taniguchi
Department of Applied Electronics, Tokyo University of Science, Japan

The technical term "nanoimprint" first appeared in "nanoimprint lithography," as used by Professor S.Y. Chou in 1995 [1]. "Nano" means 10^{-9}, and usually refers to nanometer (nm) scale objects and structures. "Imprint" means to press and make engraved marks, and so has almost the same meaning as pressing, embossing, and printing. However, lithography has a special meaning, and is the main technique for fabricating nanopatterns in the semiconductor process. The lithography process is shown in Figure 1.1.

First, a photoresist is coated on a silicon (Si) substrate. A photoresist is a material whose solubility changes when exposed to light (photons). The photomask is made of quartz and chromium (Cr), producing a light contrast – the quartz area is transparent whereas the Cr area does not transmit light. Thus, the photomask defines the area of the photoresist that will be exposed to light. An excimer laser (KrF: wavelength 248 nm, ArF: wavelength 193 nm) is used as the light source. The photomask is placed over the photoresist on Si, then light is exposed through the photomask (Figure 1.1(a)) to produce the exposed areas of the photoresist (Figure 1.1(b)). The exposed areas are changed into two types by liquid immersion. This liquid is called the developer, and the liquid immersion process is called development. After development, the photoresist where the exposed areas

Nanoimprint Technology: Nanotransfer for Thermoplastic and Photocurable Polymers, First Edition.
Edited by Jun Taniguchi, Hiroshi Ito, Jun Mizuno, and Takushi Saito.

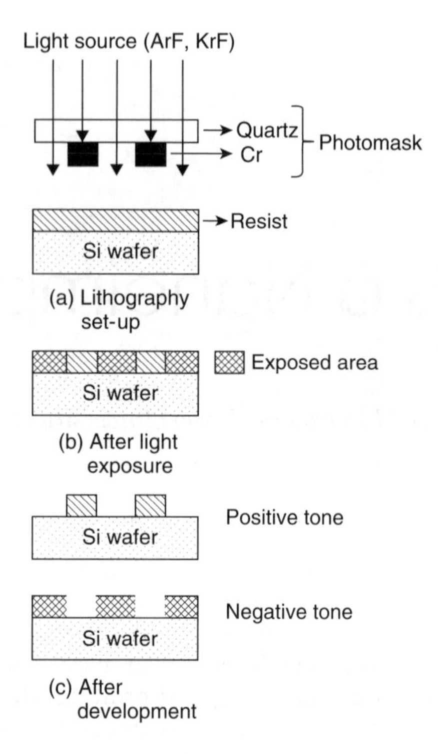

Figure 1.1 Lithography process

were removed is called positive type whereas the photoresist where the exposed areas remain is called negative type. These two types form the resist pattern on the silicon wafer. Using the resist patterns, successive semiconductor processes such as dry etching, ion implantation, and metal wiring are carried out. Dry etching is the process of removing silicon substrate using the developed resist for an etching mask. Dry etching uses an active gas such as CF_4, SF_6, or CHF_3 for the silicon substrate, by creating a plasma at low pressure. The activated species (ions or radicals) also etch the development resist, hence the term "photoresist." The ion implantation process is the process of doping donors and acceptors to create p- and n-type regions. Metal wiring is performed by the lift-off process, as follows: after development, metal is deposited by sputtering or evaporation, then the resist is removed by the remover, which dissolves the resist polymer. After removal of the resist, metal wiring remains on the silicon substrate and this area acts as an electrode and power supply. Therefore, the resolution of the resist pattern is very important for all processes because lithography determines the design rule of silicon devices such as ultra-large-scale integrated circuits (ULSIs). The design rule is the gate length or half pitch of line and space, and this index measures how small the transistor is. A small design rule enables many transistors to be formed per unit area, enabling a densely integrated electronic circuit which can be used to create high value-added devices such as large memory devices and high-performance central processing units (CPUs).

The following photolithography equation determines the resolution and hence the design rule for lithography [2]:

$$R = k_1 \times (\lambda/\mathrm{NA}) \tag{1.1}$$

where R is the resolution, k_1 is a process factor depending on the optical system of the stepper or scanner, λ is the wavelength of the light source, and NA is the numerical aperture of the lens, given by

$$\mathrm{NA} = n \times \sin\theta \tag{1.2}$$

where n is the refractive index of the light path and θ is the angle of aperture, thus $0^{\circ} < \theta < 90^{\circ}$. The photolithography exposure system includes a stepper and scanner, which can reduce the exposed pattern area to 1/4 of the mask pattern by reduced-projection optical lenses. Here, "stepper" means the "step and repeat" motion of the Si wafer stage during light exposure and "scanner" means the continuous motion of the Si wafer stage during light exposure. This system has precise stage and optical elements, and so the cost of the system is extremely high. According to eqs (1.1) and (1.2), a fine pattern can be obtained by a small wavelength (λ) and a large NA. Thus, photolithography has a limit to miniaturization; various techniques are required to exceed this limit, which are usually expensive.

In contrast, nanoimprint lithography (NIL) can exceed this limit because the patterning mechanism is merely physical pressing. The NIL process is shown in Figure 1.2.

First, a nanoscale patterned mold is prepared. A silicon wafer with resist layer is also prepared (Figure 1.2(a)). Two types of resist layer are mainly used: thermoplastic polymer and photocurable polymer. The thermoplastic polymer is solid at room temperature, but begins to flow (liquefies) upon applying heat. Thus, the shape of the thermoplastic polymer is deformed by heating and pressing of the mold. Meanwhile, the photocurable polymer is liquid at room temperature and so is easily deformed by mold pressing [3]. However, to solidify this resin, exposure to ultraviolet (UV) light is required, for which a mercury lamp i-line (365 nm) is usually used. UV light does

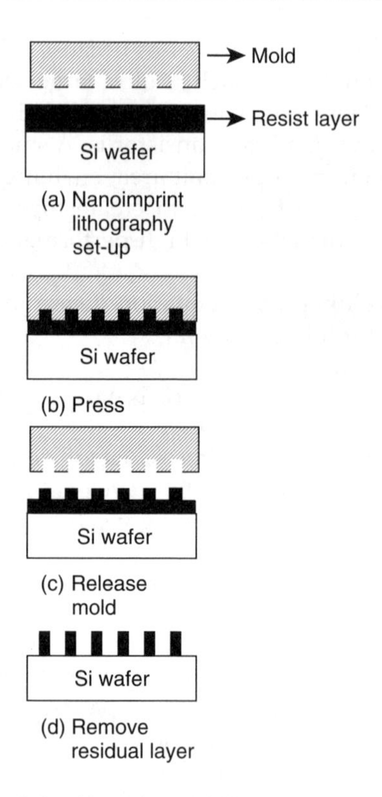

Figure 1.2 Nanoimprint lithography process

not transmit through the silicon wafer, so the mold must be UV-transparent. Quartz or sapphire is transparent to UV light, and so these materials are used for the mold. After preparing the pattern transfer, the mold presses the resist layer on the silicon wafer (Figure 1.2(b)). After solidification of the resist layer, the mold is released from it (Figure 1.2(c)). At this time, the convex part of the mold engraves the concave part of the resist layer, but the convex part of the mold does not contact the silicon wafer. Usually, a residual layer remains above the silicon wafer. This residual layer is unnecessary and is removed by oxygen plasma ashing and so on (Figure 1.2(d)). After these processes, the silicon wafer has a mask pattern as shown in Figure 1.1(c), therefore NIL can act as a lithography process.

The advantages of NIL are as follows. It is a simple process and thus cost-effective; once a nanoscale mold has been prepared, nanoresolution patterns can be obtained at low cost. Furthermore, sub-10 nm feature patterns by NIL were reported in 1997 [4], which is a major step in the semiconductor field because NIL is a simple, cost-effective, and high-resolution process. The potential of NIL is well known worldwide, and many companies and researchers are currently conducting semiconductor research [5]. In addition, NIL is very useful for other fields such as three-dimensional (3D) pattern transfer. When NIL is used for the semiconductor process, the residual layer must be removed, but in other fields it is not necessary to remove it. When the residual layer is removed, a mask pattern for silicon is obtained, but this is a two-dimensional pattern. That is, the silicon surface is painted with or without the resist mask. In contrast, by using a mold with a 3D pattern, the nanoimprint process creates a 3D replica. This kind of 3D fabrication is difficult to achieve by photolithography. Furthermore, 3D replica patterns are widely used for optical elements and surface-modified uses. For example, a moth-eye structure (which is a kind of anti-reflective structure), diffractive optical elements, gratings, Fresnel lenses, polarizers, sub-wavelength plates, and wire-grid polarizers are all optical devices. Surface-modified devices include cell culture plates, hydrophobic surfaces (lotus-effect surfaces), and adhesive surfaces such as gecko finger structures. Therefore, NIL is widely used for 3D nanofabrication, and this versatile process is called "nanoimprint technology." Therefore, NIL now means not only lithography but also 3D fabrication. This book mainly describes nanoimprint technology for 3D fabrication.

Many preparations are required to perform nanoimprint technology, such as the transfer polymer, mold fabrication process, transfer machine, and measurement system. The main transfer polymers are thermoplastic polymer and photocurable polymer, but their transfer processes are different. In addition, different transfer machines are also required for each polymer. Thus, in this book, thermoplastic and photocurable polymers are dealt with in separate chapters.

The pattern transfer of thermoplastic polymer is described in Chapters 2 and 3. First, Chapter 2 describes the history of polymer processing and the principle of the transfer method. Then, Chapter 3 describes the characteristics of thermoplastic polymer and the transfer method, and also simulation results. These simulations are very helpful for identifying thermoplastic behavior and how deformation develops over time. Nanoimprint technology using thermoplastic polymer requires a thermal cycle, so this kind of NIL is called "thermal cycle NIL" or simply "thermal NIL." The technical terms "thermoplastic polymer," "thermoplastic resin," and "thermoplastic" have almost the same meaning.

Mold fabrication processes are described in Chapter 4. The mold is the key component of nanoimprint technology, and so it is very important to be able to make a fine and precise mold. The mechanical cutting process and electron beam lithography and dry etching process required to obtain a nanoscale 3D shape mold are described in detail. Machine tools and accurate machine positioning and control have been developed, enabling sub-micrometer order 3D cutting shapes to be fabricated. The merit of using cutting tools is rapid fabrication. Electron beam lithography (EBL) involves exposure to an electron beam instead of excimer laser light. An electron beam can be focused to less than several nanometers, so a finer pattern (less than 10 nm) can be delineated. Electron beam lithography is usually used for the photomask in the semiconductor process, but by using EBL and successive dry etching technologies, nanoimprint molds can be fabricated. This book also describes various mold materials. The technical terms "mold," "stamp," and "template" have almost the same meaning.

Nanoimprint technology using photocurable polymer is described in Chapter 5. In this case, ultraviolet light is used to harden the photocurable polymer, so this kind of NIL is called "ultraviolet NIL," or UV-NIL. This chapter describes the UV-NIL mechanism, photocurable polymer science, UV-NIL machine, release agents, and measurement methods. Usually, nanoscale patterns are observed with a scanning electron microscope (SEM) and atomic force microscope (AFM), but these methods are for microscale local observation. In this chapter, a macroscale non-uniform measurement system is described. The release agent is the coating material on the mold surface, which prevents the photocurable polymer from sticking. The technical terms "photocurable polymer," "photocurable resin," "UV-curable polymer," and "resin" have almost the same meaning.

Chapter 6 outlines the latest nanoimprint technologies, as well as actual applications and some devices made by nanoimprint technology.

This book outlines nanoimprint technology using thermoplastic and photocurable polymers, and describes in detail nanoscale transfer technology, materials, machines, know-how, and trends.

References

[1] Chou, S.Y., Krauss, P.R., and Renstrom, P.J. 1995. Imprint of sub-25 nm vias and trenches in polymers. *Appl. Phys. Lett.* **67**: 3114–3116.

[2] Mack, C. 2007. *Fundamental Principles of Optical Lithography*. John Wiley, Chichester, UK, pp. 21–22.

[3] Haisma, J., Verheijen, M., van den Heuvel, K., and van den Berg, J. 1996. Mold-assisted nanolithography: A process for reliable pattern replication. *J. Vac. Sci. Technol. B* **14**: 4124–4128.

[4] Chou, S.Y., Krauss, P.R., Zhang, W., Guo, L., and Zhuang, L. 1997. Sub-10 nm imprint lithography and applications. *J. Vac. Sci. Technol. B* **15**: 2897–2904.
[5] Colburn, M., Johnson, S., Stewart, M., Damle, S., Bailey, T., Choi, B. *et al.* 1999. Step and flash imprint lithography: A new approach to high-resolution patterning. SPIE 24th International Symposium on Microlithography: Emerging Lithographic Technologies III, Santa Clara, CA, pp. 379–389.

2

Nanoimprint Lithography: Background and Related Techniques

Hiroshi Ito[a] and Takushi Saito[b]

[a]*Department of Polymer Science and Engineering, Yamagata University, Japan*

[b]*Department of Mechanical and Control Engineering, Tokyo Institute of Technology, Japan*

2.1 History of Material Processing: Polymer Processing

Throughout human history, appliances have been produced using various materials. Very early devices were made of wood or stone, but metal came to be used after several thousand years. We have passed through a long period of development, and our current civilization involves the construction of tall, rigid bridges and skyscrapers using metal materials for their frames.

In contrast, the use of polymer materials began with natural rubber, progressing to natural cellulose, with dramatic development over a very

Nanoimprint Technology: Nanotransfer for Thermoplastic and Photocurable Polymers, First Edition.
Edited by Jun Taniguchi, Hiroshi Ito, Jun Mizuno, and Takushi Saito.

short timeframe. The use of synthetic resins has enabled the production of indispensable components and products that are important to modern society. Such wide use is generally attributable to the superior molding properties of processed polymers. The degrees of freedom in product shape and coloration for such materials are high. In addition, the strength ratios (material strength divided by density) of materials classified as engineering plastics approach those of metal materials. Increasingly, the value of these characteristics is recognized for use in containers, housing, household electrical appliances, and interior panels of cars as well as their bumpers. Moreover, mechanical parts and optical components are made less and less with metal materials and optical glass materials; they are instead made with polymer materials for reasons of performance, machined strength, and optical characteristic.

A number of techniques are used to give shape to polymer materials, but the shaping technology called plastic molding is used most often industrially. The basic concept of plastic molding is divisible into three steps, as follows.

Step 1: Fluidization process for polymer materials
Thermosetting and UV-curing resins are liquid, but the thermoplastic resin melts on heating. Materials must be fluid to enable shaping in the next step.

Step 2: Replication process for polymer materials
Polymer materials are molded to shape with a metal die and stamper. When the polymer materials have high viscosity, the materials should flow. Therefore, high pressure must be applied to materials to shape the mold. Furthermore, the effects of interfacial force are greater for smaller replication shapes.

Step 3: Heating/cooling and demolding processes for polymer materials
As with heated thermosetting resin, cooled thermoplastic resin attains a molded shape and UV resin cures with ultraviolet irradiation. During such processes, stress occurs in the product. It often happens that the product shape changes and optical characteristics are lost.

In practice there will be some overlap in the processing steps described above, although polymer processing can generally be divided into three steps. For example, we perform nanoimprinting in one device during steps 1–3, and injection molding during steps 2 and 3, which progress simultaneously inside a die. Close attention to the various parts of the process is required, but such an overlap realizes high productivity of the plastic molding process. The characteristics of the molding technique are explained in detail in other chapters.

2.2 Products with Microstructure and Nanostructure

From the Industrial Revolution in the 18th century up to the mid-20th century, manufacturing processes were undertaken on a millimeter scale. From the latter half of the 20th century to the beginning of the 21st century, that operational scale has been reduced to micrometers and nanometers (Figure 2.1).

The equipment supporting the growth of such miniaturization technology includes electron microscopes and atomic force microscopes. These modes of microscopy were invented in the first half of the 20th century. Subsequently, applications of the equipment rapidly became popular. Using such equipment, the microscopic world has been revealed from the sub-micrometer to the nanometer scale to a degree that was impossible using optical microscopy alone. For example, a certain virus might be only a few hundred nanometers long, despite inflicting great damage on people. Researchers would have been entirely unable to observe a virus on this scale using optical microscopy.

Some products with a fine surface structure on the micrometer to nanometer scale have been produced; their use in many fields is greatly anticipated. In the medical field, regenerative medical techniques have been proposed using tissue scaffolds with a fine nanostructure [1, 2]. Physiological sensors using surface plasmons have been produced [3]. Moreover, in the semiconductor and electronic component manufacturing fields, development of a transistor, memory, and sensor using miniaturization and carbon nanotubes as wiring for CPUs is progressing [4]. New patterned media for high-density memory for data storage are also being considered [5]. Furthermore, a nanostructure on the surface of a film with advanced features of an electrolyte membrane has been produced for use in the energy field of polymer electrolyte fuel cells (PEFCs) and direct-methanol fuel cells (DMFCs) [6, 7]. These products not only have microscale and nanoscale surface structures, they also offer additional value and performance.

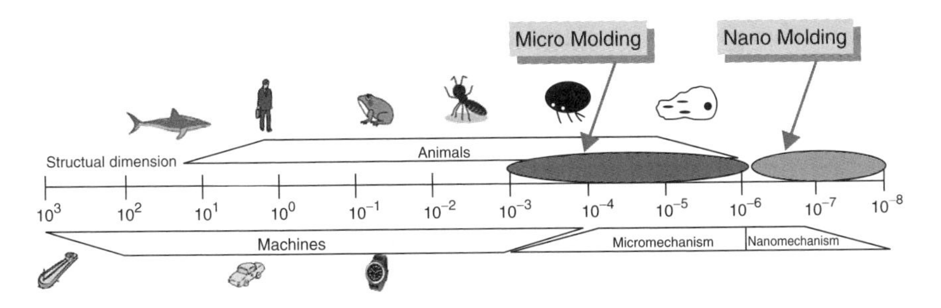

Figure 2.1 Schematics of structure dimension and molding scale

For example, it is necessary to give organism affinity to tissue scaffold materials. Electronic components and a fuel cell require thermal resistance and stability during a chemical reaction. Consequently, in these fields, coexistence of product shape and functional characteristics is important. Furthermore, mass production can be realized to reduce the costs of manufacture.

Moreover, memory and CPUs with semiconducting integrated circuits are described as the way to advance the miniaturization of structures. In the International Technology Roadmap for Semiconductors (ITRS) [8], dynamic random access memory (DRAM) of 1/2 pitch is set as the line width standard of semiconductor integrated circuits serving at the 68 nm level in 2007 and around 25 nm in 2015. In this field, ArF liquid immersion exposure technology, using an advanced mode of photolithography, is proposed using an ArF excimer laser as a light source with 193 nm oscillation wavelength [9]. The mode improves resolution using a liquid with a high refractive index as a filling between the substrates used as objective and target. Moreover, as a phase shift method, the phase and hardness of the light which passes a mask are changed to form a phase shift mask and a photomask by preparing the portion for which refractive index and permeability differ [10]. Thereby, resolution is improved. For super-ultraviolet radiation with extreme ultraviolet (E-UV) and soft X-rays of wavelength 13.5 nm light source exposure technologies [11, 12], the resolution at the time of exposure itself is also improved. Using these technologies, mass production conversion of the pattern formation at 45–30 nm level has begun. Consequently, products with a microstructure or nanostructure are always in demand for miniaturization and functional improvement of the fine structure. To achieve miniaturization, top-down construction methods are important which produce the desired shape using a processing machine with a mold stamper. Also increasingly important are bottom-up techniques using materials' own self-organization.

2.3 Technology for Making Micro- and Nanostructures

Several techniques are used to fabricate products of micro- and nano-size structure. For example, a photolithography process is used in the manufacture of semiconductor integrated circuits (Figure 2.2). In this technique, exposure light is irradiated on the photoresist material spread on the silicone substrate through the mask, carving circuit patterns. The difference in callousness of the photoresist material caused by the exposure light irradiation forms a base of minute patterns on the substrate. Then, actual minute patterns are formed by removing the unnecessary photoresist material and performing an etching process on the substrate surface. Because this technique can efficiently create a microscale structure of large area, it has contributed to

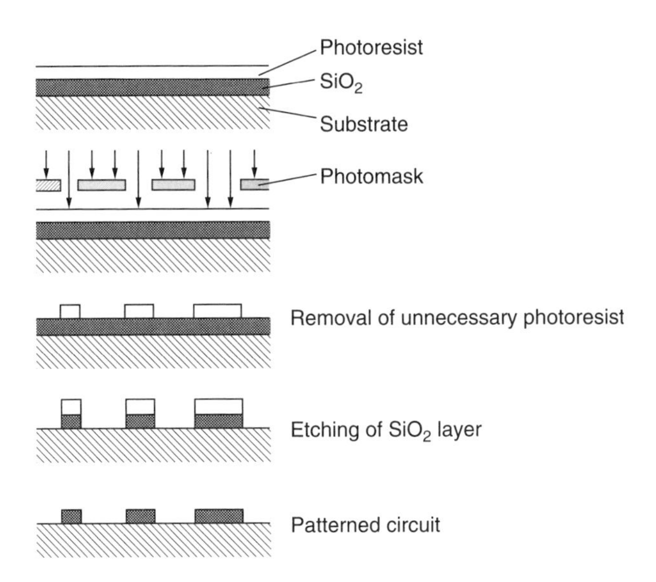

Figure 2.2 Schematics of the photolithography process

mass production and a reduction in the cost of semiconductor manufacture. However, the facility cost of the process is very high. Moreover, various contrivances to shorten the wavelength of the light source are needed to create smaller circuits, because the spatial resolution of the circuit pattern on the substrate depends on the wavelength of exposure light.

The photolithography process is also used as a method of making a minute structure and a movable mechanism in MEMS (micro-electro-mechanical systems). During the fabrication steps of MEMS, a removal process is needed not only in the plane direction of the substrate but also in the depth direction. Note that the ratio between the size in depth and plane directions is called the aspect ratio. Deep RIE (reactive ion etching) using ICP (inductively coupled plasma) is used to make a structure with a high aspect ratio (structural size in the depth direction higher than that in the plane direction).

The "LIGA process" which makes a minute pattern by irradiating synchrotron radiation or X-rays on the resist film thickly painted on the substrate is also well known as a technique for making minute structures with high aspect ratio [13]. In recent years a great reduction in process cost has been achieved, because a new resist film material (for example, SU-8™) that can use ultraviolet light as exposure light source has appeared [14]. In contrast, there are other techniques which directly fabricate a minute structure on the substrate. For example, EBs (electron beams) and FIBs (focused ion beams)

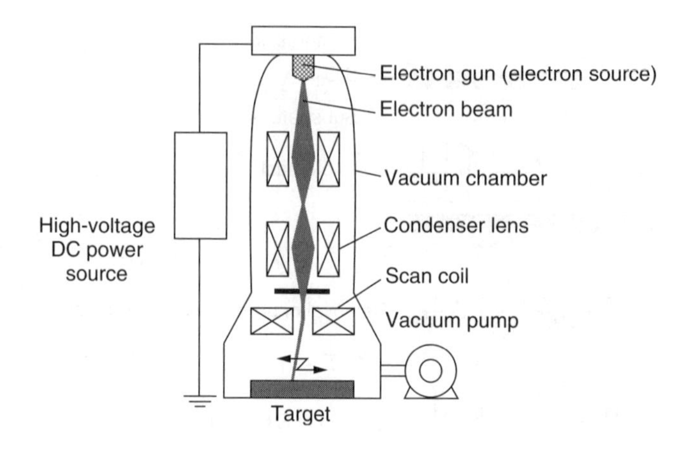

Figure 2.3 Structure of the electron beam scanning system

are often used to directly carve minute structures (Figure 2.3). In these techniques, if the optical system for beam focusing is appropriate, the spatial resolution of the process can reach one-digit nano size. However, it usually takes a significant amount of time to fabricate the structure over a large area because the process basically progresses a line at a time.

Microfabrication by EB and FIB is categorized as the removal process, because electrically accelerated electrons and ions collide at high speed and remove the material. A similar removal process can be realized using an ultrashort pulse laser with high energy. However, it is known that light energy has superior advantages in microfabrication. For instance, the fabrication of a minute statue was reported using UV curing material [15]. A key technique was use of the multiphoton absorption process of a femtosecond laser. In recent years miniaturization of machining tools has also been advanced, and a micro-end mill tens of micrometers in diameter is being marketed. To achieve highly accurate microfabrication by machining, total system development including the machine tool and its driving device has become important. Owing to the size limitation of tools, the minimum scale of current processing is thought to be around several micrometers. However, because an increase in the number of tooling axes results in a processing degree of freedom, more complex shapes can be achieved.

Although the techniques described above have merits and demerits, they are mostly technically established. However, the idea of a process that makes an object directly and individually is inefficient as a technique for mass production. Therefore, the process of a mother pattern made by the techniques explained above is indispensable for the mass production of products with micro- and nanostructure. In such circumstances, an injection molding method and nanoprint/imprint methods considered as the replication process are regarded as promising techniques suitable for mass production.

References

[1] Yang, F., Murugan, R., Ramakrishna, S., Wang, X., Ma, Y.-X., and Wang, S. 2004. Fabrication of nano-structured porous PLLA scaffold intended for nerve tissue engineering. *Biomater.* **25**: 1891–1900.

[2] Smith, L.A., Liu, X., and Ma, P.X. 2008. Tissue engineering with nano-fibrous scaffolds. *Soft Matter* **4**(11): 2144–2149.

[3] Weilbaecher, C.R., Hossain, M., Gangopadhyay, S., and Grant, S. 2007. *Development of a novel nanomaterial-based optical platform for a protease biosensor*. Proc. SPIE, Vol. 6759.

[4] Naeemi, A. and Meindl, J.D. 2009. Carbon nanotube interconnects. *Ann. Rev. Mater. Res.* **39**: 255–275.

[5] Nanostructured materials in information storage. *MRS Bull.* **33**: 831–834.

[6] Zhang, Y., Lu, J., Shimano, S., Zhou, H., and Maeda, R. 2007. Nanoimprint of proton exchange membrane for MEMS-based fuel cell application. Proc. 6th Int. IEEE Conf. on Polymers and Adhesives in Microelectronics and Photonics, pp. 91–95.

[7] Zhang, Y., Lu, J., Wang, Q., Takahashi, M., Itoh, T., and Maeda, R. 2009. Nanoimprint of polymer electrolyte membrane for micro direct methanol fuel cell application. *ECS Trans. Micro Power Sources* **16**: 11–17.

[8] http://www.itrs.net/

[9] Honda, T., Kishikawa, Y., Tokita, T., Ohsawa, H., Kawashima, M., Ohkubo, A. *et al.* 2004. *ArF immersion lithography: Critical optical issues*. Proc. SPIE, Vol. 5377, Part 1, pp. 319–328.

[10] Martinsson, H., Sandstrom, T., Bleeker, A., and Hintersteiner, J.D. 2005. Current status of optical maskless lithography. *J. Microlith., Microfab., Microsyst.* **4**: 1–15.

[11] Tawarayama, K., Aoyama, H., Magoshi, S., Tanaka, Y., Shirai, S., and Tanaka, H. 2009. *Recent progress of EUV full field exposure tool in Selete*. Proc. SPIE, Vol. 7271.

[12] Smith, H.I., Carter, D.J.D., Ferrera, J., Gil, D., Goodberlet, J., Hastings, J.T. *et al.* 2000. Soft X-rays for deep sub-100 nm lithography, with and without masks. *Proc. Mater. Res. Soc. Symp.* **584**: 11–21.

[13] Malek, C.K. and Saile, V. 2004. Applications of LIGA technology to precision manufacturing of high-aspect-ratio micro-components and -systems: A review. *Microelectron. J.* **35**: 131–143.

[14] Lu, B., Xie, S.-Q., Wan, J., Yang, R., Shu, Z., Qu, X.-P. *et al.* 2009. Applications of nanoimprint lithography for biochemical and nanophotonic structures using SU-8. *Int. J. Nanosci.* **8**: 151–155.

[15] Kawata, S., Sun, H.-B., Tanaka, T., and Takada, K. 2001. Finer features for functional microdevices. *Nature* **412**(6848): 697–698.

3

Nanopattern Transfer Technology of Thermoplastic Materials

Takushi Saito[a] and Hiroshi Ito[b]
[a]*Department of Mechanical and Control Engineering, Tokyo Institute of Technology, Japan*
[b]*Department of Polymer Science and Engineering, Yamagata University, Japan*

3.1 Behavior of Thermoplastic Materials

3.1.1 Thermoplastics

Many plastic products exist in the world. Plastics are strictly defined as thermoplastics. However today, as one of the various polymer materials including thermosets, "plastics" might refer to the plastic molding and its surroundings. Synthetic resins are polymer materials intended for molding. They are classifiable into thermoplastic and thermoset resins. In addition, according to the illumination used for resin curing, some are designated as UV-curing resins used in the field of semiconductor manufacture. Treatment-resistant layer (negative resist) resin is one such material. Thermoplastic

Nanoimprint Technology: Nanotransfer for Thermoplastic and Photocurable Polymers, First Edition.
Edited by Jun Taniguchi, Hiroshi Ito, Jun Mizuno, and Takushi Saito.

resin (a crystalline polymer in which crystallization occurs) and amorphous polymer (in which no crystallization occurs) are two categories. Table 3.1 presents a typical plastic material. In general, the crystalline resin has a glass-transition temperature and a melting temperature (or melting point). The melting temperature is defined by the crystal melting temperature. Above this melting point, the crystalline resin has a solid–liquid phase transition. It will be completely liquid at a higher temperature than the melting temperature. In injection molding, thermoplastic is poured into a mold in a liquid state, where it solidifies and cools below the melting point.

Table 3.1 Typical plastics

Thermoset	Phenol-formaldehyde (PF) Urea-formaldehyde (UF) Melamine-formaldehyde (MF) Unsaturated polyester resin (UP) Epoxy resin (EP) Polyurethane (PUR) Polyimide (PI) ...		
Thermoplastic	General plastics	High-density polyethylene (HDPE)	crystalline
		Low-density polyethylene (LDPE)	crystalline
		Polypropylene (PP)	crystalline
		Polystyrene (PS)	amorphous
		ABS resin (ABS)	amorphous
		Acryl resin (PMMA) ...	amorphous
	Engineering plastics	Polyamide (PA)	crystalline
		Polyacetals (POM)	crystalline
		Polycarbonate (PC)	crystalline
		Polybutylene terephthalate (PBT)	crystalline
		Polyethylene terephthalate (PET) ...	crystalline
	Super engineering plastics	Polysulfone (PSU)	amorphous
		Polyether ether ketone (PEEK)	crystalline
		Polyphenylene sulfide (PPS)	crystalline
		Polytetrafluoroethylene (PTFE) ...	crystalline

It then reproduces the moldings. For amorphous material, the melting point is not evident because of the lack of crystallization; only the glass transition temperature T_g can be present. The molecular chain mobility changes at this temperature, from a glass (solid state) to a rubber (slightly soft state) above T_g. At temperatures higher than T_g it changes from a rubber to a liquid, and the processability is improved. For amorphous materials, the product will be obtained by pouring into a mold at a temperature well above T_g, with subsequent cooling at a temperature less than T_g.

In thermoplastics, plastics with higher heat deflection temperature and strength are designated as engineering plastics; they are used as industrial plastics. In engineering plastics, plastics with higher heat deflection temperature are designated as super engineering plastics. These have a very high melting point and glass transition temperature, generally near 300 °C.

In contrast, thermosetting resin forms a polymer network structure by polymerization caused by heating. The resin cannot revert to a liquid after curing; if the resin is heated further, then burning results. When thermosets are molded, a low molecular weight resin having liquidity is molded into a predetermined shape; then the product will be obtained by reaction because of heating, etc. The adhesive is also thermosetting; in this case, a curing agent is used as a common reagent. A UV-curing resin is an illustrative example of a photocurable resin. This resin changes chemically from a liquid to a solid in response to UV light energy. Processes such as printing, coating, and painting are possible with a UV-curing resin. In the field of nanoimprinting, new UV/photocurable resin materials have been developed.

3.1.2 Basis of Viscoelasticity and Rheology

Polymers are long strings of more than 10,000 molecular weight called polymer chains. Therefore, polymers have a "spaghetti"-like structure which becomes tangled in a form resembling that of long noodles. Their physical properties display great complexity. In particular, their flow and deflection properties differ from those of low molecular weight materials. The study of the flow and deformation of materials is called "rheology." As one example of a rheologically studied material, polymer is a "viscoelastic" material that combines viscosity and elasticity as a property. The examination of viscoelastic polymers constitutes a large area of polymer rheology studies. This investigation has contributed not only to the development of very basic science linking the movement and orientation of polymer chains and rheological properties, but also to the development of high-performance polymer materials and their evaluation.

To understand viscoelastic properties, which are actually mechanical properties, it is simplest and most convenient to examine the response to an object's deformation. The rheological properties of a polymer, its viscoelastic properties, depend on the temperature and loading rate. For example, polymers can be bent using little power in a short time by warming with a heater or dryer. It is also well known empirically that polymers can be bent by applying a weight to the polymer over a long time. This characteristic is known as the temperature–time superposition principle (or temperature–frequency superposition principle), which is very important in the field of rheology.

3.1.3 Measurement of Rheology

Measurements include static and dynamic methods. Dynamic methods generally measure according to the temporal state. In particular, dynamic sine vibration is added in most cases. In contrast, static measurements show constant motion. Therefore, the measurement of a liquid polymer is designated as a steady flow measurement. Thixotropic (isokinetic lifting technique) measurements are designated as static measurements, while dynamic viscoelastic measurements are simply designated as sine vibration measurements. In a solid, tensile, creep, and stress relaxation tests are classified as static measurements. Sine vibration measurements are defined as dynamic viscoelastic measurements, as in a fluid. Many fixtures are also possible; for the measurement of liquid polymers, there are jigs with cylindrical, double-disk cones (cone–plate), parallel disks (plate–plate), etc. Solids are classified using measurements such as tensile, bending (one support or two supports), and torsion measurements.

Rheological measurements are used to determine the relation between force and strain. Therefore, strain control measures the force necessary for deformation. Stress control measures how the deformation will be loaded. Evaluations of liquid polymers generally include measurements of the shear rate γ (s^{-1}) and shear stress σ (Pa) in static measurements, the angular velocity ω (rad/s), and the complex modulus G^*, or storage modulus G' (Pa) and loss modulus G'' (Pa), decomposed from G^* in the dynamic measurement. If the shear stress σ increases linearly with the increasing shear rate, it is a Newtonian fluid. Consequently, the flow curve of the relation between the shear rate and stress is linear through the origin in a Newtonian fluid. However, the most common polymer fluid cannot pass through the origin and is also nonlinear. Such fluids are called non-Newtonian fluids. Bingham fluid is one fluid which does not flow until the critical stress, and appears as a Newtonian fluid above that stress. In the general polymer fluid, viscosity decreases rapidly with increasing shear rate, which is designated as shear thinning (shear softening). Figure 3.1 presents the relationship between shear viscosity (Pa • s = shear stress/shear rate) and shear rate.

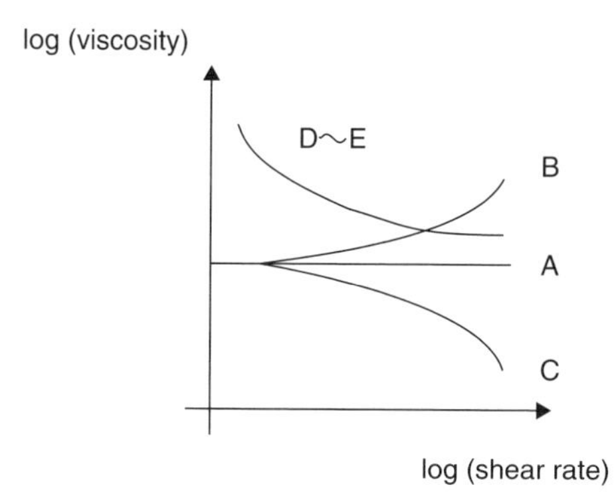

A, Newtonian fluid; B, dilatant fluid; C, thixotropic fluid (such as polymer melt or solution); D–E, fluid with yield stress (e.g., Bingham fluid).

Figure 3.1 Relation between the shear rate and the viscosity of various fluids (logarithmic)

The storage modulus G' is an elastic component that is thought to result from such structures as coil vibration and agglomeration in the polymer. The loss modulus G'' is a viscous component that is equivalent to static shear stress. Consequently, the loss modulus G'' divided by the angular velocity ω is called the dynamic viscosity η', which can be thought of as equivalent to the static viscosity η. Meanwhile, the temperature on the vertical axis and the storage modulus E (Pa), the loss modulus E'' (Pa), tan δ and the loss tangent (E''/E') on the horizontal axis are thought of in terms of a solid.

When the melting viscoelastic properties are measured using a rheometer, the rheological properties of polymers from the solid to the molten state are understood by plotting roughly three times the storage modulus G' and the loss modulus G'' against E' and E''. This graph can show, qualitatively, Young's modulus at each temperature from E' along with the softening temperature, crosslinking density, and flow temperature from that change. The glass transition temperature and entangled state of the polymer chain can also be measured qualitatively from E'' or tan δ, along with the miscibility and crystallinity for that shape.

3.1.4 Physical Properties of Viscoelastic Materials and the Temperature–Time Superposition Principle

For a polymer that is randomly entangled with each molecular chain and mixed in the amorphous and crystal state, the temperature–frequency dependence is determined according to the viscoelastic properties of molecules. Molecular chains of these materials have different exercise modes, releasing energy in a specific frequency and temperature for each mode. The damping performance will be the largest movement in the transition region, which becomes the active molecule. Local molecular motion of the main chain in the glass area will be transformed into micro-Brownian motion of the main chain in the transition area, which gives the greatest losses. The Williams–Landel–Ferry (WLF) equation, which is the temperature–frequency superposition principle, proved this transition region expression. Here, an important basic concept is that it can represent similar elastic properties for time because it replaces temperature and frequency (time) based on the temperature dependence of the elastic modulus of a viscoelastic material. Figure 3.2 presents the temperature dependence of a viscoelastic material. The horizontal axis shows logarithmic time. The vertical axis shows the logarithmic relaxation modulus $E(t)$. In the figure, the left-hand side shows data measured at each temperature $T_1, \ldots, T_7$; the right-hand side shows the curve created based on these, designated as the master curve.

First, it explains data from the left side: T_1 is the cold side; T_7 is the hot side. The relaxation curve of the modulus obtained from about 10 to 1000 s on the horizontal scale is shown. The viscoelastic behavior of a polymer depends on the space available to exercise heat inside the molecular chains freely (free volume). In the low-temperature region, a polymer does not show elastic behavior because of the small free volume; this condition is designated as the glassy state. In Figure 3.2, T_1 corresponds to this state. In this case, $E(t)$ indicates an approximately constant value of 10^9 Pa. The elastic modulus of polymers in the glassy state is generally about this value.

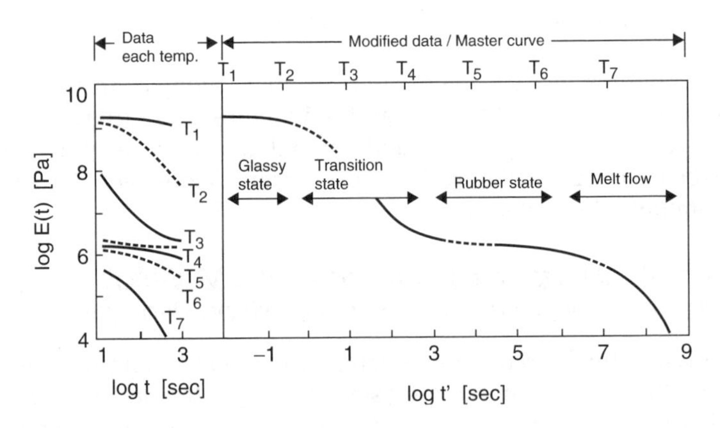

Figure 3.2 Simple master curve of the material

The free volume increases with higher temperature. The molecular chains will be able to exercise while receiving viscous resistance; hence viscoelastic behavior appears. This condition is designated as the rubber state. Here, $T_4, \ldots, T_6$ correspond to the rubber state. The elastic modulus is about 1 to 1000 times lower than for the glassy state in this example. Polymers have a glass transition point, which is the material-specific temperature at which viscoelasticity is expressed, that is, the transition from glass to rubber state. It is used as an index representing the nature of the solid polymer. The glass transition temperature is near T_2, at which point the curves start to decline. Here, the stress–relaxation curves measured at different temperatures move along the horizontal axis (time axis). A single curve is designated as the master curve. The right-hand side of Figure 3.2 shows the results used to create the master curve. It indicates that the curve at T_3 was chosen as the basis from the curves depicted on the left of Figure 3.2, arranged so that one curve is attributable to shifting the other curves in the horizontal axis. The horizontal axis of the obtained curve covers a wide span of time: 0.01–10^9 s (30 years). This figure suggests that the behavior from 0.01 s to 30 years can be estimated using the results of 10 to 1000 s at several temperatures. The horizontal axis in the figure represents logarithmic time. Consequently, operations to shift the curve to the horizontal axis correspond to multiplying each curve by the ratio of time α_T, as follows:

$$\alpha_T = t/t' \tag{3.1}$$

In this equation, α_T denotes the shift factor (temperature–time factor or move conversion factor), t stands for the time at any temperature T (K), and t' signifies the time at the reference temperature T_0 (K). The deformation mechanism of thermally rheologically simple materials has the same temperature dependence at all time regions. When the temperature changes from T_0 to T, the relaxation time is uniformly α_T times T_0. The WLF equation and the Arrhenius equation are known as typical temperature–time superposition principles.

The WLF equation is a relation between α_T and T. The following is generally used:

$$\log \alpha_T = -\frac{C_1(T - T_R)}{C_2 + (T - T_R)} \tag{3.2}$$

Here, when T_R is about 50 K higher than the glass transition temperature, C_1 and C_2 are, respectively, 8.86 and 101.6 K, not related to the materials. The glass transition temperature is the temperature to transit from the temperature region of elastic response to the temperature region of viscoelastic response, which is generally used as the index reflecting the attenuation characteristics of the resin. Although C_1 and C_2 are known as constants with

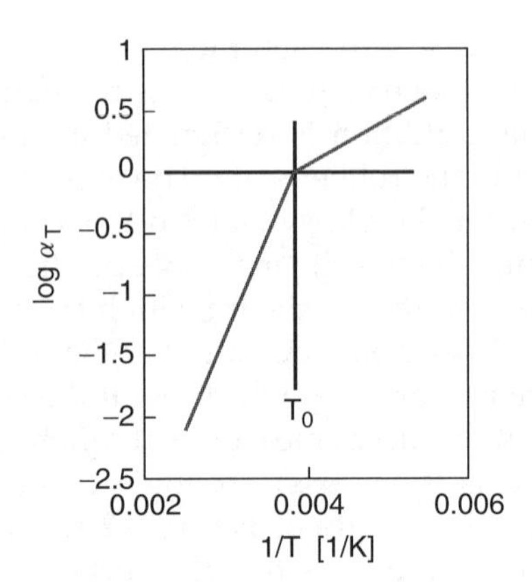

Figure 3.3 Example of the temperature-time superposition principle (Arrhenius equation ($T_g = 260$ K))

value greater than that for a typical polymer not mixed with filler, they are recognized as a guide only.

The Arrhenius equation applies the concept of chemical reaction kinetics to the process of viscous flow:

$$\log \alpha_T = \beta \frac{\Delta H}{R} \left(\frac{1}{T} - \frac{1}{T_0} \right) \tag{3.3}$$

Here, ΔH signifies the activation energy [J/(mol K)], R stands for the gas constant (1.98×10^{-3} kcal/mol K), T (K) denotes temperature, and T_0 (K) represents the reference temperature. Figure 3.3 shows the relationship between α_T and temperature when T_0 in eq. (3.3) is 260 K. As the form of the equation clarifies, eq. (3.3) presents a linear relation by taking the inverse of the temperature on the horizontal axis.

3.1.5 Materials Design for Realizing Nanoimprints

Lowering the target material viscosity is important to pour the material into the microstructure region. Consequently, a perfect temperature exists to reduce viscosity. Additionally, the molecular weight distribution must generally be reduced because of the low molecular weight components. However, the strength and elastic modulus decrease because of the low molecular weight.

It is possible to predict the viscosity and temperature from the temperature–time superposition principle described above. This can serve as a guideline for the optimization process. Additionally, experimentation has revealed that a sufficiently high modulus must reduce fracturing of the structure when it is demolded. Consequently, sufficiently lower viscosity and solidification must hold for heat imprinting of thermoplastics. Moreover, it is important that the difference in modulus of elasticity for glass and rubber regions and the flow area be larger. Currently, it is necessary for chemical manufacturers to develop new materials to provide these properties. Additionally, control of the wettability against materials is important for the mold, surface tension, and interfacial tension. These properties are relevant material parameters that are useful in conjunction with rheological properties.

3.2 Applicable Processes Used for Nanopattern Transfer

3.2.1 Introduction of Injection Molding Process

In industry, many kinds of polymer materials are currently being used. The reason for this is quite simple: chemical companies manufacture various kinds of material according to user needs because the features and physical properties of polymer materials differ significantly.

Polymer materials are classified roughly into thermoplastic and thermosetting materials (optically curable material is an example of the latter). At the molecular level, the heating of thermoplastic material corresponds to an increase in thermal vibration of molecules having one-dimensional structure. Therefore, the degree of freedom of the intertwined polymer molecules increases due to heating, and a molten state results. In contrast, the molecules of a thermosetting material unite chemically with other molecules by heating and form a three-dimensional structure. Therefore, the relative movement of the molecules is restrained, and the transformation of the entire material becomes small. The thermosetting material does not soften after it stiffens once, even if it is heated again. In contrast, thermoplastic material can soften and harden many times based on its temperature.

This feature of thermoplastic material, to control the molten phase and solid phase according to temperature, enables the various types of processing method (the thermal imprinting process being one of them). Before our main discourse explaining the thermal imprinting process, the injection molding process which can make a molded product with microstructure is explained here (see Figure 3.4).

First, a few millimeters of granular polymeric material is placed into the plasticizing unit of an injection molding machine. The screw rotating in the unit melts and mixes the material until a uniform molten polymer is obtained. This is called the melting process, and it is an important process in which the physical properties of the molded product are determined. Although the

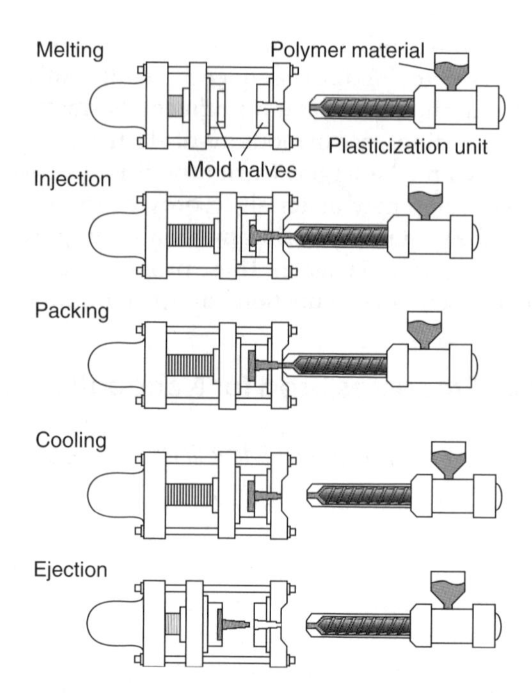

Figure 3.4 Process steps of injection molding

melt temperature is different for each material, it is usually set between 200 and 300 °C, and often becomes about 400 °C in the case of super engineering plastics. However, the melt temperature described here is different from the glass transition temperature and the crystallization temperature, both of which are physical properties of a polymer material. Briefly, the melt temperature in the injection molding process is the temperature of the state in which the molten polymer can be injected at any time (state with sufficient fluidity). This temperature, of course, is higher than both the glass transition temperature and the crystallization temperature.

Subsequently, the injection process that fills the mold cavity with the molten polymer is started. In general, the metal mold for the injection molding is divided into two halves (a fixed half and a movable half). Molten polymer is injected into the space of the mold that is formed by closing these halves. The temperature of the mold is usually maintained from room temperature to about 100 °C. To obtain molded products of good quality, it is preferable that the mold temperature is set equal to the melt temperature, and the molded material is cooled over a long period of time. However, productivity is the most important issue during actual production.

Therefore, it is important that the temperature of the mold be set far lower than that of the molten polymer in order to cool down the molded material quickly. As a result, the surface of the molten polymer flowing in the mold cavity is cooled rapidly by touching the metal mold surface. The problems caused by this are described in detail later.

It is convenient to adopt the velocity of the flow front as an index that represents the speed of the molten polymer flow in the mold cavity, because the independently measured physical properties of the molten polymer can be compared with the common index. However, the flow front speed of the molten polymer in the actual mold depends on its position, because a general molded product has a complex three-dimensional geometry. Therefore, the injection rate (cm^3/s) based on the extruded volume of molten polymer from the nozzle of the plasticizing unit is often used as index. The molten polymer that flows in a narrow cavity of the mold causes a large pressure drop. The pressure setting of the injection molding machine must exceed this pressure drop, and is called the injection pressure.

In actual injection molding, the filling of molten polymer continues after the completion of the filling stage. This secondary filling is needed to supplement a decrease in the volume caused by the cooling shrinkage of the material in the mold. This step is called the packing process, and it is important because the size accuracy of the molded product is thereby improved efficiently. The pressure setting of the injection molding machine is called the packing pressure, and is generally set higher than the injection pressure described above. The time to maintain the packing pressure is called the holding time. By adopting the packing process, the residual stress remaining in the final product usually increases because the material cooled down in the mold is fluidized compulsorily. Therefore, adoption of the packing process often has a bad influence on optical characteristics in the manufacture of optical components.

Next is the cooling process, which occupies a large portion of the cycle time of the injection molding process. The cooling period depends on the kind of polymer material and the size of the molded product. In the case of a short period, it ends in a few seconds. In contrast, the cooling period might take one minute or more when it is long. It is preferable to maintain the cooling process until the temperature of the molded polymer decreases enough to stabilize the quality of the products. However, if the surface temperature of the molded polymer falls below the glass transition temperature or the crystallization temperature where the product can maintain its shape as the molded product, the cooling step usually ends. By adopting this kind of actual technique, high productivity of the injection molding process can be achieved.

A series of injection molding processes ends by removing the molded product from the mold. Note that this state is an early stage for the next molding, and the same processes are repeated for continuous molding. If necessary, the molded product is processed in a secondary process (such as coloring and joining), to become a final product.

As described here, the method of injection molding efficiently combines the processes of giving shape to the molten polymer and cooling of the molded material. Therefore, the injection molding process is recognized as a powerful mass-production tool in which uniform quality can be realized over a wide range of sizes and with a high shape degree of freedom. As is generally known, this process is widely used to manufacture molded products from car bumpers (meter size) to cellular phone connectors (millimeter size). Based on the same idea, the examination of focusing mass production of the microstructure (micrometer size) on the injection molding process is being actively pursued [1–3].

3.2.2 Problems of the Injection Molding Process

In the injection molding process, the filling and cooling of the polymer material progress simultaneously because molten polymer of high temperature is injected into a mold cavity of low temperature (Figure 3.5). Therefore, the region of low temperature that is called the surface solidification layer is formed in the surface part that comes into contact with the mold surface. Generation of the surface solidification layer has a great influence on the injection situation of molten polymer in the injection molding process. Also, the existence of the surface solidification layer greatly influences the shape accuracy and function of the molded products.

Because the surface solidification layer is generated even in the injection process, it is difficult to transcribe a microstructure on the mold surface

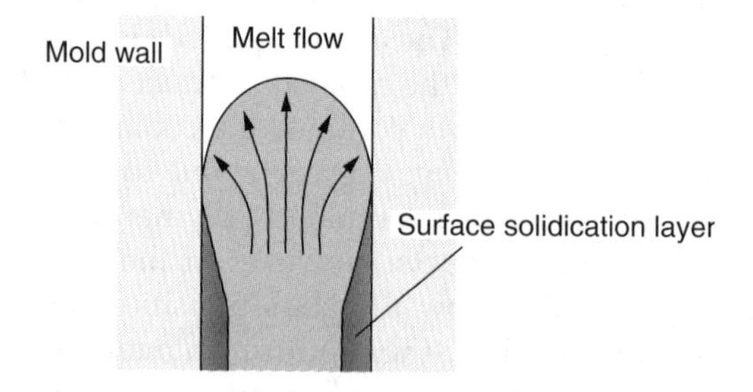

Figure 3.5 Molten polymer flow in the cavity and the incidence of surface solidification layer

completely. When the thickness of the surface solidification layer is large enough against the cavity gap, a large pressure drop is caused in the filling stage. In the extreme case, filling of the polymer material becomes imperfect (called a short shot). In addition, polymer molecules at the boundary between the surface solidification layer and the main current are highly stretched in the flow direction, and this causes molecular orientation. Because the molecular orientation causes mechanical and optical anisotropy, it often ruins the quality of precision parts and optical components. Therefore, controlling the growth of the surface solidification layer in the injection process becomes important to improve the quality of the molded products. To address this, there is a technique that actively changes the mold temperature in one process cycle [4–7]; the temperature of the mold is raised at the filling stage, and decreased at the cooling stage.

Because the fluidity of molten polymer is important in the injection molding process, the melt temperature of the process is often set higher than the thermal imprint temperature. It is well known that the glass transition temperature of PMMA is about 100 °C. And the melt temperature in the injection molding process is set from 200 to 220 °C while the thermal imprint temperature is set from 130 to 150 °C. The temperature difference between the two significantly affects the material shrinkage and thermal deformation of the obtained products. In the injection molding process that adopts a higher temperature, the cooling of the material causes larger thermal shrinkage. The size accuracy of the molded products can be improved by adopting the packing process. However, even if the packing process is adopted, transcription of a minute structure on the molded surface is not improved completely because the surface solidification layer obstructs the pushing effect of the packing pressure. Moreover, adoption of the packing process might increase the residual stress and deformation. It is necessary to note that this kind of problem ruins the profile accuracy and optical quality of the molded products.

3.2.3 Advantages of the Thermal Imprinting Process

As described above, a molded product with a complex three-dimensional geometry can be obtained by using the injection molding process. In contrast, the shape of the mother stamp with a microstructure arrayed in two dimensions is transcribed on the heated sheet or film by the thermal imprinting process. During the thermal imprinting process, the transcription of a microstructure progresses under isothermal conditions, and the application of pressure to the processing material is relatively low. Therefore, the rate of strain and the amount of deformation given to the material are small, and the deformation inevitably caused in the process can be reduced quickly due to the isothermal condition. Owing to these characteristics, it can be said that the thermal imprinting process is more advantageous than the injection molding process for the following reasons: higher transcription of microstructure, low residual deformation, and low residual molecular orientation. Moreover, this process can result in a product with a microstructure on a thin substrate (note that this kind of product cannot be made easily by the injection molding process). However, it should be noted that a microstructure must have a straight shape along the stroke direction of the stamp. Moreover, the thermal shrinkage difference between the polymer material and the stamp becomes a significant problem when the size of the transcription region of the microstructure is enlarged to increase the throughput. In such a case, exfoliation of the product without anticipation might occur, and the transcribed microstructure is destroyed.

The injection molding process is advantageous for making products with complex geometry at a high production rate. Therefore, the mass production of minute products is being pursued by miniaturizing the injection molding machine. However, the surface solidification layer inevitably generated in the injection molding process causes a significant problem when a minute shape of micrometer order is transcribed. To solve this problem, it is necessary to divide the process of transcription of a minute shape in the polymer material and the cooling process of the polymer material. From this point of view, a thermal imprinting process is thought to be reasonable.

3.3 Pattern Transfer Mechanism of Thermal Cycle NIL

3.3.1 Introduction of Thermal Imprinting Process

Chou *et al.*'s research paper [8] is well known as an initial work on the thermal imprinting process. Because the thermal imprinting process can use various thermoplastic polymer materials and glass as process material, it has the potential for making products with high added value (such as optical elements, mass storage devices, and disposable medical treatment inspection kits).

The outline of this process is shown schematically in Figure 3.6. First of all, the polymer material of the substrate is heated to more than the temperature required to transcribe the shape of the microstructure. Then, the stamp used to carve the microstructure being transcribed is pushed against the substrate. A pressure of several megapascals is often adopted to improve the accuracy of the transcription shape. Because this compression step is included in the thermal imprinting process, the stamp must have constant strength although the polymer material has been softened by the temperature

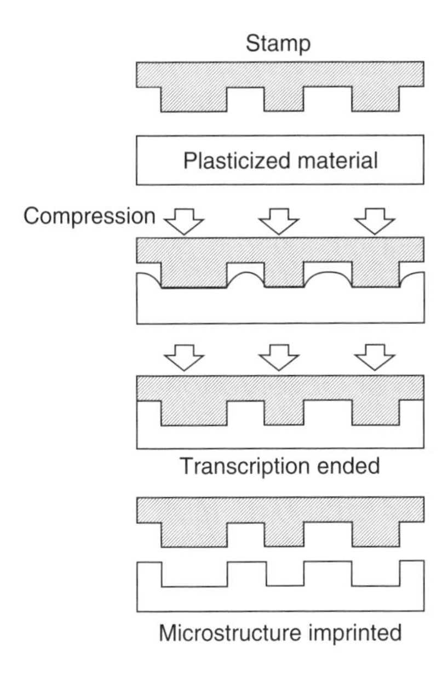

Figure 3.6 Steps of the thermal imprinting process

rise. Therefore, silicon, silicon carbide, glassy carbon, and tantalum are often used as stamp material. An electroplated nickel mold as used for micro-injection molding can also be employed. After transcription, the molds are cooled down to the temperature at which the polymer material solidifies and can be removed. Therefore, shortening of the heating and cooling cycle becomes an important factor for the enhancement of productivity. However, there is a limit to the speed of heating and cooling because the mold has a large thermal capacity. To solve this difficulty, several possibilities have been tested; such as reconsideration of the heat exchange system of the machine, and an increase in the amount of production per process time by enlarging the substrate size. However, destruction of the transcription pattern may occur, especially in peripheral parts, because the shrinkage difference on cooling between the polymer material and the stamp is enlarged with enlarging the substrate size. Moreover, removing the product from the stamp becomes difficult when high temperature and pressure are adopted to improve the transcription accuracy. Although coating of the mold release agent on the stamp surface is often adopted as a solution for this problem [9], there is a possibility that the existence of the release agent will influence the material behavior in case of extremely minute structure transcription.

3.3.2 In-situ Observation of Thermal Imprinting Process

In the thermal imprinting process, the material is treated as a viscous fluid or viscoelastic fluid. Although many papers [10–16] have revealed the behavior of the polymer material in microscale transcription via offline observation and model analysis, there are few investigations that directly observe the material behavior during the process. Thus, an experimental device that enables in-situ observation of the process was constructed, and some results of the transcription behavior obtained by the device are described here.

3.3.2.1 Construction of Observation System

Figure 3.7 shows the thermal imprint device and the microstructure fabricated on the stamp surface. In the test section, the stamp with the microstructure was fixed on a lower plate. A window for the observation and the fixture device for the processing material were fixed on an upper plate. Small electric heaters were installed in both upper and lower plates. In the experiment, the microstructure on the stamp was transcribed to the substrate of the polymer material by pushing the lower plate driven by the piezo actuator (maximum stroke of 55 μm). A digital microscope system equipped with a long-focus lens was used to observe the transcription behavior.

The stamp used in the experiment had minute straight grooves, 15 μm wide and 5 μm deep on a silicone substrate. In other words, the polymer material flow in a minute groove, 15 μm wide was observed in the experiment. The polymer material used in the experiment was an extruded plate of poly(methyl methacrylate). The sample was dried sufficiently before the test.

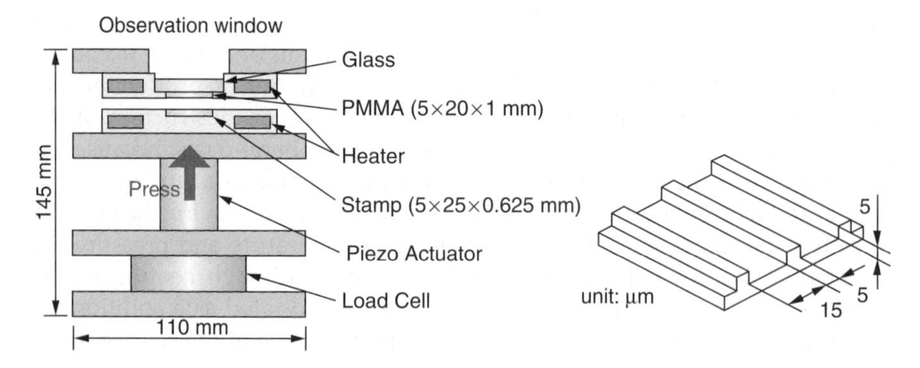

Figure 3.7 Thermal imprint device and microstructure on the stamp surface

An exploratory experiment by differential scanning calorimetry showed that the glass transition temperature of the material was 103 °C.

3.3.2.2 Observation Result of Minute Transcription

Figure 3.8(a) shows the observed result for a temperature of 130 °C and pressure of 0.5 MPa. The dark-colored region indicates the part where the stamp surface came into contact with the surface of the polymer material, and the light-colored region indicates the part where a gap existed between the stamp and the polymer material. In this condition of relatively low process temperature, the contact region of the polymer material and the stamp progressed first according to the longer direction of the groove, and then gradually extended in the direction of the width of the groove with the passage of time. The results show that the microstructure existing on the stamp surface influenced the transcription condition as well as the general molding process, similar to the injection molding process.

Next, the observed result at a temperature of 200 °C and pressure of 0.5 MPa is shown in Figure 3.8(b). The time change of the contact region (dark-colored part) between the polymer material and the stamp was greatly different from that of the previous result, shown in Figure 3.8(a). That is, the transcription of the microstructure was initiated from a part where the surface of the polymer material came into contact with the stamp. Then, the contact region (transcription) extended homogeneously from the starting point. This result was due to the fact that the transcription of the polymer material was improved because the material viscosity decreased with the rise in temperature.

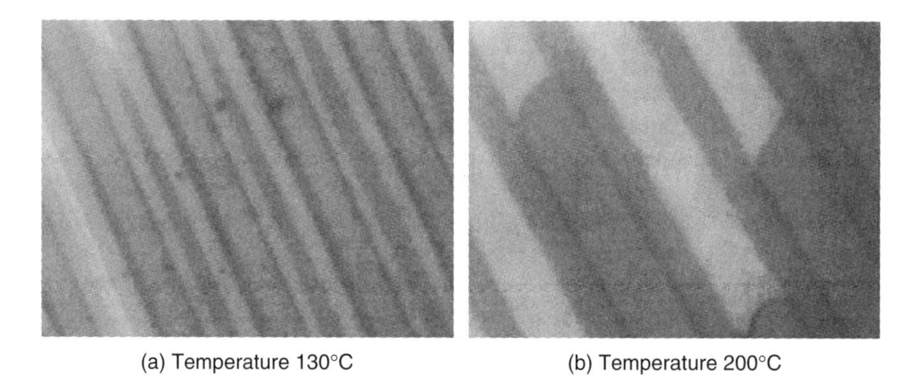

(a) Temperature 130°C (b) Temperature 200°C

Figure 3.8 Magnified image of the polymer substrate transcription

To evaluate the influence of the process temperature on the transcription behavior, image processing was performed on the obtained images. The ratio of the area that became dark (contact area between the polymer material and the stamp) was obtained through processing, and this ratio is regarded as an indicator of the transcription progress. Figure 3.9 shows the time-dependent transcription progress at the different process temperatures, with pressure of 0.5 MPa in this case. The horizontal axis of the graph represents logarithmic time, and the time required for material transcription was greatly shortened by the rise in process temperature. For the scale of microstructure used here, material transcription was mostly complete at 5 s in the case of the process temperature being about 150 °C.

Although the evaluation of the material behavior described above concerns the overall transcription shape, more detailed examination is needed to discuss sub-micrometer scale transcription, such as the corner shape of the microstructure. Thus, a laser confocal microscope was used to evaluate the transcription condition, and the result is shown in Figure 3.10. The evaluation values here are the radius of the circular arc L that touches the residual gap at the corner and the width Wu of the transcription top. The curved geometry of about 0.5 μm in the corner radius was found by preliminary measurement of the stamp of the initial state, though the width of the stamp groove part was almost 15 μm according to the designed size. By considering these results, it was shown that an almost complete transcription was achieved with the condition of the process temperature exceeding 200 °C. Note that this temperature is considerably higher than the usual processing temperature of 130–150 °C.

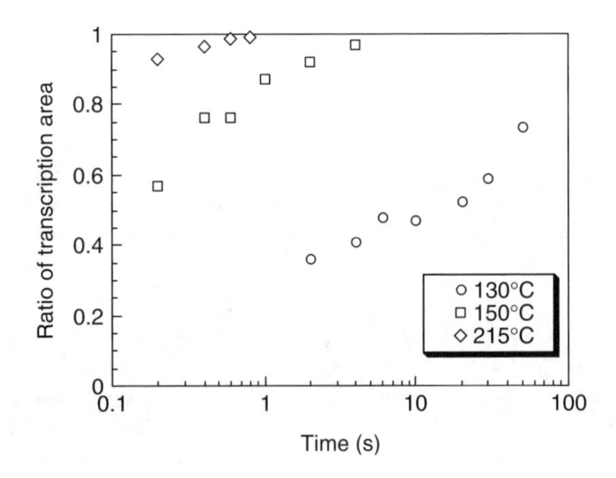

Figure 3.9 Time-dependent transcription progress at different process temperatures

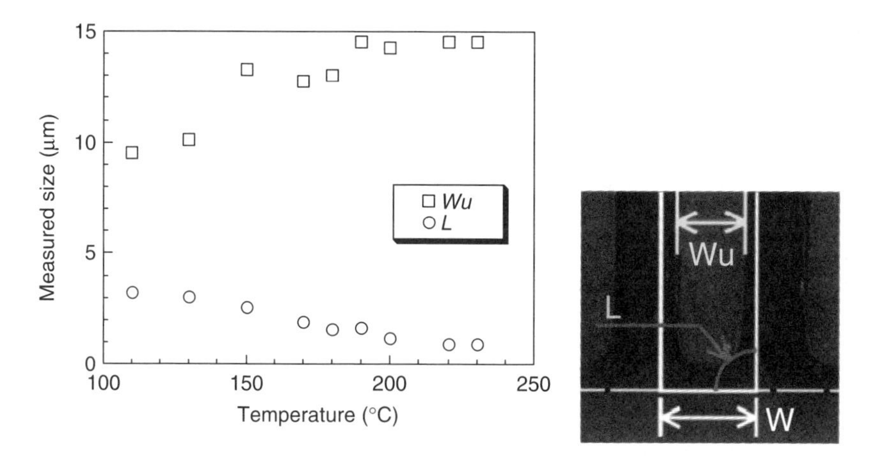

Figure 3.10 More precise evaluation of the transcription shape

3.3.3 Offline Measurement of Replication Process in Thermal Cycle NIL

Recently, thermal nanoimprint lithography (TNIL) techniques have been reported [17–26]. TNIL is a similar process to hot embossing techniques, and has been well known and applied to development nanopatterns and commercial productions for micro and nano devices. However, nano- and micro-replication of the mold stamper is complex, and it is difficult to control the replication properties under optimum imprinting conditions. As the basis of these problems, online measurement of replication properties was conducted recently. However, the online measurement of replication was still performed on a microscale structure because the measurement technique used optical devices. Therefore, offline measurement of the nanoscale-structure replication of the mold is important. This technique measures the change in replication rate over time. The replication surface is observed using a confocal laser scanning microscope, atomic force microscope, and scanning electron microscope, etc. A question mark remains over the accuracy of the absolute value due to the influence of the deformation of molded sheets or films at demolding and the elastic shrinkage of film or sheet materials at thermal history. However, this technique is very easy and it excels in promoting understanding of qualitative replication behavior.

In order to clarify the relationship between material properties and replication properties in thermal nanoimprinting, two different kinds of polycarbonate (PC) sheet were imprinted at various conditions. In this study, two different grades of PC sheet (Iupilon S-3000N and H-3000, supplied by Mitsubishi Engineering-Plastics Corporation) were used as imprinting materials. The sheet thickness was 400 μm. Iupilon S-3000N and H-3000 sheets show 15 and 30 g/min of melt mass flow rate, respectively. H-3000 could have a relatively lower molecular weight and melt viscosity than S-3000N. As-cast PC sheets were almost isotropic. Nanoimprinting tests were performed using thermal nanoimprinting equipment (NanoimPro type515, Nanonics Corporation). The imprinting scheme is illustrated in Figure 3.11. An embossing mold and a polymer sheet were inserted between the mirror surface board and the shock absorber. Before inserting the embossing mold and the polymer sheet, static electricity was removed to prevent contamination. The imprinting temperature, pressure, and processing time were controlled independently. The imprinting conditions in this study are summarized in Table 3.2. The condition 0 is a basic imprint condition, imprinting at 165 °C and 5 MPa for 60 s. Compared with the condition 0, the imprinting pressure was increased from 5 MPa to 8 MPa and 12 MPa in conditions 1 and 2. The imprinting time was varied from 60 s to 30 s and 90 s in conditions 3 and 4. The imprinting temperature was also increased from 165 °C to 175 °C in condition 5. After imprinting, the PC sheet was kept with the pressure applied until the temperature dropped below the glass transition temperature of 125 °C naturally without heating.

After imprinting tests, the replicated pattern was characterized. The height of the imprinting pattern was measured using a confocal laser scanning microscope (VK-9700, Keyence). Figure 3.12 shows SEM images of the mold stamper and replicated sheets. The nanopattern on the sheet surface was observed. Figure 3.13 indicates the replicated height by imprinting with the various conditions in Table 3.2. The height of the embossing mold pattern

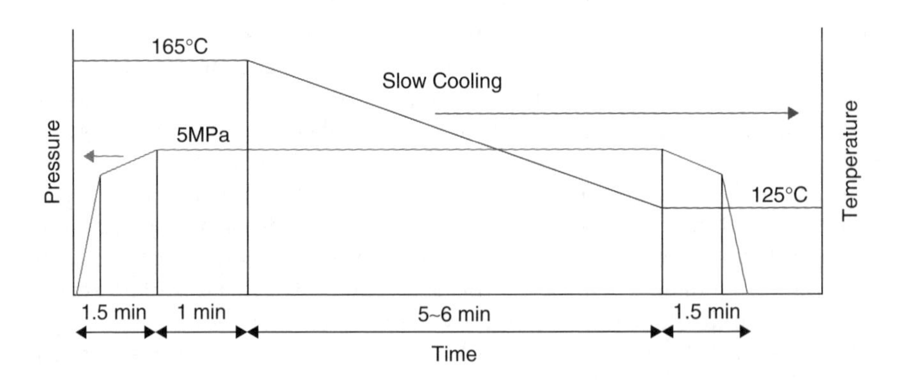

Figure 3.11 Process scheme of thermal nanoimprinting

Table 3.2 Thermal imprinting conditions

Condition number	Imprinting temperature (°C)	Imprinting pressure (MPa)	Imprinting time (s)
0	165	5	60
1	165	8	60
2	165	12	60
3	165	5	30
4	165	5	90
5	175	5	60

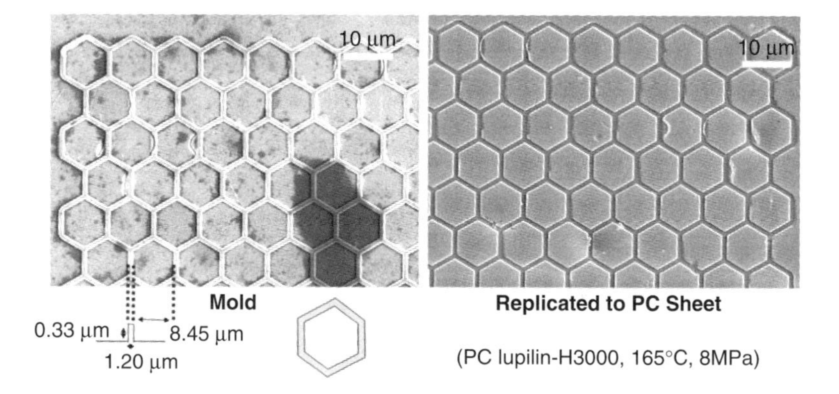

Figure 3.12 Photograph of imprint mold and molded sheet

was 330 nm, which is shown as a broken line in Figure 3.13. In the basic condition 0, the imprinted height was around 260 to 270 nm, which is 80% of the replicated height of the embossing mold pattern. Compared with condition 0, a higher imprinting pressure was applied in conditions 1 and 2. The imprinted height was around 280 to 310 nm. The replicated height was improved compared with condition 0. The precise difference in replication properties between S-3000N and H-3000 was not confirmed in conditions 0, 1, and 2. In conditions 3, 4, and 5, the imprinting time and temperature were varied. The imprinted height for S-3000N sheets was around 260 to 300 nm, which was equivalent to conditions 0, 1, and 2. In the case of H-3000 sheets, the imprinted height exceeded 330 nm, which was the height of the embossing mold. These results show that the replication property was particularly affected by the imprinting time and temperature.

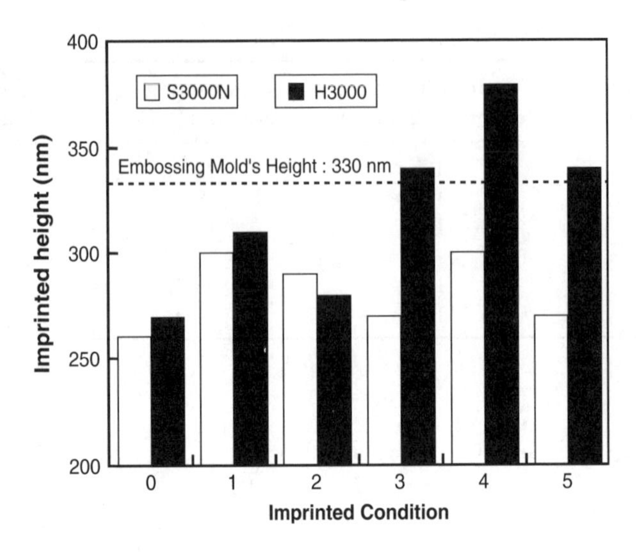

Figure 3.13 Replicated heights by thermal imprinting for S-3000N and H-3000 sheets with various imprinting conditions as listed in Table 3.2

3.4 Modeling of Nanopattern Transfer

3.4.1 Importance of Viscosity in Thermal Imprinting Process

In the thermal imprinting process, a stamp having microstructure on the surface is mechanically pushed against a substrate of thermoplastic material. From the microscale viewpoint, it can be said that the rigid structure of the stamp penetrates into the substrate of the softened polymeric material. The progress of the penetration is affected not only by the viscosity of the polymeric material, but also by the structural geometry and density of the stamp. Hirai *et al.* [27, 28] reported that the resolution in the replication of thermal imprint is dominated by the aspect ratio of the surface structure of the stamp and the initial thickness of the substrate of polymeric material. In this section, the reason for this is explained in terms of the viscosity of thermoplastic material.

As described in Section 3.1.4, the stiffness of thermoplastic material lessens when the temperature rises. Therefore, the use of a higher temperature in the thermal imprinting process is efficient and reasonable to transcribe the minute structure of the stamp surface. However, there is an upper limit of temperature that is known as the thermal degradation point of the material. Furthermore, the capillary bridge shown schematically in Figure 3.14(a)

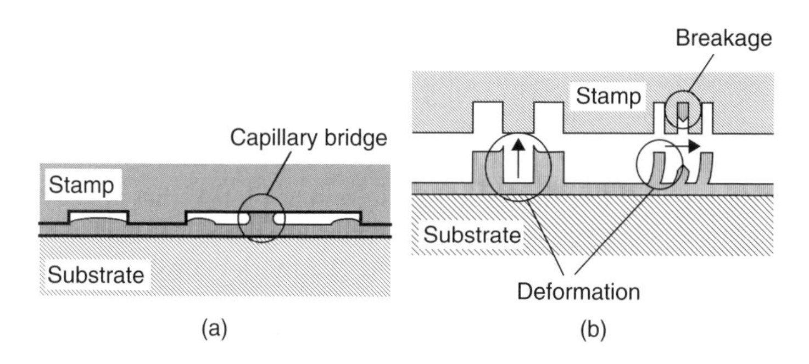

Figure 3.14 Defects in the thermal imprinting process

is generated spontaneously and tends to lower the transcription accuracy [29, 30] if the viscosity of the thermoplastic material is extremely low. Moreover, the difference in thermal expansion ratio between the stamp and the polymeric material causes a large thermal stress at the interface when the high temperature condition is applied in the process. As shown in Figure 3.14(b), this kind of stress often brings about deformation or breakage of the transcribed structure at the demolding stage.

The viscosity of a molten polymer depends on the deformation rate, as explained in Section 3.1.3. The viscosity of a molten polymer generally decreases with increase in shear rate from the zero-shear viscosity, and this is known as the shear-thinning phenomenon. Therefore, not only the zero-shear viscosity but also the effective viscosity has to be considered in the actual thermal imprinting process. The temperature distribution in the thermal imprinting process is relatively homogeneous, and this is a difference from the injection molding process. In the thermal imprinting process, however, considerable distribution of the effective viscosity is caused by the local distribution of the flow velocity of the molten polymer [31]. This means that the transcription progress depends on the position of the contact interface even if the imprint device pushes the stamp against the polymeric material at constant force.

The driving force of the transcription in the thermal imprint is the pressure acting on the stamp in contact with the substrate of polymeric material. Thus, it is expected that the application of higher pressure tends to enhance the flow of polymer melt and improve the transcription progress. As an example, the residual layer thickness is important not only in the thermal imprint but also in the UV imprint as discussed here. A squeeze flow is caused in the residual layer by the descent of the stamp against the substrate. By simplifying the situation, the force required to generate the squeeze flow between two circular plates arranged in parallel is described as follows [32]:

$$F = \frac{\pi \ r^4 \eta v}{8a^3} C \tag{3.4}$$

where r, η, v, a, C are the radius of the circular plates, the viscosity of the fluid, the approaching speed of the two parallel plates, the half value of the distance between the parallel plates, and the model constant, respectively. The important point being indicated in the equation is that the force required to generate the squeeze flow is rapidly increased by decreasing the distance between the circular plates. For this reason, it is quite difficult to realize a residual layer thickness close to 0.

Schift and Heyderman [33] proposed that if a constant imprint force F per length is applied to the protrusion part of the stamp, then the following equation gives the variation of the residual layer thickness h of the line-shaped stamp:

$$\frac{1}{h^2(t)} = \frac{1}{h_0^2} + \frac{2F}{\eta_0 s^2} t \tag{3.5}$$

where h_0, η_0, s, t are the initial layer thickness of the polymeric material, the zero shear viscosity, protrusion width, and time, respectively. This equation indicates that the protrusion width s has a great influence on the required time for thinning of the residual layer thickness. The model equations introduced here suggest that the transcription progress of the thermal imprint is essentially dominated by the viscosity of the polymeric material.

When a process of thermal imprinting is designed, the glass transition temperature particularly for amorphous materials is regarded as an important index to determine the imprint temperature and the demolding temperature. Shift *et al.* [34] suggested that a temperature 50–70 °C higher than the glass transition temperature is appropriate for the imprinting process, and a temperature 20 °C lower than the glass transition temperature is desirable for the demolding process.

As described above, the viscosity of the polymeric material dominates the progress of the thermal imprinting process. The viscosity of the material can be measured using a rheometer. However, the data obtained using a rheometer is the value under ideal conditions. In the actual process, important conditions for the imprinting process – such as temperature, shear rate, and pressure – depend on the time and position. Thus, it is not easy to determine suitable conditions that can realize good transcription in the thermal imprinting process. A numerical simulation will be an important tool in solving this kind of difficulty. Estimation at the size of micrometers or nanometers might be realizable by utilizing the simulation well. Moreover, information regarding the processing window in an actual process could be obtained by calculating the distributions of the temperature and pressure in the entire stamp. The first step in numerical simulation of the imprinting process is explained in the following section.

3.4.2 Mathematical Treatment in Injection Molding and Thermal Imprinting Process

Replication for micro- and nanopatterns is a complex phenomenon in relation to flow behavior in the filling or transcription stage, shrinkage in the cooling stage, deformation at mold release, etc. Numerical simulations are cost-effective and provide quick solutions, and thus are an essential tool for the design and evaluation of the processing parameters. Numerical investigations are able to estimate aspects of the physical model which otherwise would be difficult to quantify.

The cycle of micro- and nanosurface pattern transfer processes such as injection molding and thermal imprinting can be described as being made up of three main stages: filling, packing/holding, and cooling. During the initial filling stage, when the molten polymer fills the cavity, compressibility effects are small and the flow is usually considered incompressible because of isothermal states. Once the entire fine surface pattern is filled, the process enters the packing/holding stage during which pressure is applied on the material. At this point compression occurs and the solution behavior is driven by the polymer compressibility. In the injection molding process, the compression phase is very short but results in dramatic changes in the nature of the polymer flow in the cavity. Velocities may take values comparable with those during the filling stage, while the pressure increase propagates into the entire cavity. As soon as the pressure reaches its maximum value, the velocity amplitude falls. It is the beginning of the holding stage during which compressibility effects are still important. The inflow of material must compensate for the decrease in specific volume from cooling. The polymer in the gate then freezes and the mass of polymer in the cavity is set. The physics during the remaining cooling phase are modeled by the energy and state equations. In contrast, in the thermal imprinting process, the compression phase is longer than that of the injection molding process. The compression phase is isothermal. After compression, the resin cools down below the glass transition temperature of solidification for a polymer.

The rheological response of polymer melts is generally non-Newtonian and non-isothermal with the position of the moving flow front in injection molding [35]. Because of these inherent factors, it is difficult to analyze the filling process. Therefore, simplifications are usually used. For example, in the traditional middle-plane model and dual-domain model [36], the Hele–Shaw approximation [37] is used. So, both of these models are 2.5D models. In the 2.5D model, the velocity and variation of pressure in the gap-wise direction are neglected except that the temperature is solved by FDM, and the filling of a mold cavity becomes a 2D problem in the flow direction and a 1D problem in the gap-wise direction. As most of the injection molded parts have a sheet-like geometry in which the thickness is much smaller than the other dimensions of the part, these models have generally been successful in predicting the advancement of melt fronts, pressure fields, and temperature distribution. The interest in 3D simulation of injection molding has increased tremendously and some progress has been made in the past few years [38, 39]. One reason is the processing of large and complex parts. With the development of molding techniques, more and more molded parts have thick or non-uniform thickness, such as those with micro- and nanosurface features. In these cases, the velocity and changes of parameters in the gap-wise direction are considerable and cannot be neglected. A 3D simulation model should be able to generate complementary and more detailed information relating to the flow characteristics and stress distributions in molded parts.

In a 3D model, since the change of physical quantities is not neglected in the gap-wise direction, the momentum equations are much more complex than those in a 2.5D model. It is impossible to obtain the velocity–pressure relation by integrating the momentum equations in the gap-wise direction, which is done in a 2.5D model. The momentum equations must first be discretized, and then the relation between velocity and pressure is derived from the equations.

The pressure of the polymer melt is not very high when the cavity is being filled, so the melt is considered incompressible. Inertia and gravitation are neglected compared with viscous force. With the above approximation, the governing equations, expressed in Cartesian coordinates, are as follows.

Momentum equations:

$$\frac{\partial}{\partial x}\left(2\eta\frac{\partial u}{\partial x}\right)+\frac{\partial}{\partial y}\left[\eta\left(\frac{\partial v}{\partial x}+\frac{\partial u}{\partial y}\right)\right]+\frac{\partial}{\partial z}\left[\eta\left(\frac{\partial w}{\partial x}+\frac{\partial u}{\partial z}\right)\right]-\frac{\partial(P)}{\partial x}=0 \quad (3.6)$$

$$\frac{\partial}{\partial x}\left[\eta\left(\frac{\partial v}{\partial x}+\frac{\partial u}{\partial y}\right)\right]+\frac{\partial}{\partial y}\left(2\eta\frac{\partial v}{\partial y}\right)+\frac{\partial}{\partial z}\left[\eta\left(\frac{\partial w}{\partial y}+\frac{\partial v}{\partial z}\right)\right]-\frac{\partial(P)}{\partial y}=0 \quad (3.7)$$

$$\frac{\partial}{\partial x}\left[\eta\left(\frac{\partial w}{\partial x}+\frac{\partial u}{\partial z}\right)\right]+\frac{\partial}{\partial y}\left[\eta\left(\frac{\partial v}{\partial z}+\frac{\partial w}{\partial y}\right)\right]+\frac{\partial}{\partial z}\left(2\eta\frac{\partial w}{\partial z}\right)-\frac{\partial(P)}{\partial z}=0 \quad (3.8)$$

Continuity equation:

$$\frac{\partial u}{\partial x} + \frac{\partial v}{\partial y} + \frac{\partial w}{\partial z} = 0 \tag{3.9}$$

Energy equation:

$$\rho C_p \left(\frac{\partial T}{\partial t} + u\frac{\partial T}{\partial x} + v\frac{\partial T}{\partial y} + w\frac{\partial T}{\partial z} \right) = \frac{\partial}{\partial x}\left(K\frac{\partial T}{\partial x}\right) + \frac{\partial}{\partial y}\left(K\frac{\partial T}{\partial y}\right) + \frac{\partial}{\partial z}\left(K\frac{\partial T}{\partial z}\right) + \eta\dot{\gamma}^2 \tag{3.10}$$

where x, y, z are 3D coordinates and u, v, w are the velocity components in the x, y, z directions. P, T, ρ, C_p, K, and η denote pressure, temperature, density, specific heat, thermal conductivity, and viscosity, respectively.

The Cross viscosity model has been used for the simulation:

$$\eta = \frac{\eta_0(T,\ P)}{1 + (\eta_0\dot{\gamma}/\tau^*)^{1-n}} \tag{3.11}$$

where $\eta, \dot{\gamma}, \tau^*$ are the non-Newtonian index, shear rate, and material constant, respectively.

Because there is no notable change in the scope of melt temperature during filling, the Arrhenius model [40] and WLF model for η_0 are employed as follows.

Arrhenius type:

$$\eta_0(T,P) = B \exp\left(\frac{T_b}{T}\right) \exp(\beta P) \tag{3.12}$$

WLF type:

$$\eta_0(T,P) = D \exp\left[-\frac{C_1\left(T - T(P)^*\right)}{C_2(P) + (T - T(P)^*)} \right] \tag{3.13}$$

where B, C_1, C_2, D, T_b, β are material constants.

The momentum equations are discretized using Galerkin's method with bilinear velocity–pressure formulation [41]. The element equations are assembled in the conventional manner to form the discretized global momentum equations. The element pressure equations are also assembled in the conventional manner to form the global pressure equations. After the pressure field has been obtained, the velocity values are updated using a new pressure field because the velocity field obtained by solving the momentum equations does not satisfy the continuity equation. The boundary condition is very important to solve the flow fields in these simulations. In the cavity wall, the no-slip boundary conditions are employed.

3.4.3 Process Simulation in Micro- and Nanopattern Transfer

Numerical analysis is very effective when we understand the justification for experimental results. Moreover, it is required to improve thermal nanoimprinting and also required to elucidate various physical phenomena such as replication and internal structure in molding processes because nanomolding processes are so complicated. From such a point of view, it is very important to reveal the mechanism of replication and internal higher-order structure, such as birefringence of nano-molded surface features. Replication especially is a complex phenomenon in relation to flow behavior in the filling stage, shrinkage in the cooling stage, deformation at mold release, etc.

The modeling of the micro- and nanosurface transfer process requires time dependency in the process. Scant effort has been made in the development of numerical methods capable of handling time-dependent flow of integral constitutive equations [42–44]. To perform realistic modeling, a fully 3D as well as free surface viscoelastic flow will be required. The method of Wapperom and Keunings [43] and that of Rasmussen [44] both apply Lagrangian particle variables. Therefore, both methods can handle (large) movement of moving surfaces or interfaces without extra effort. The movement can be free or specified. Only the approach of Rasmussen has been numerically formulated in 3D, although the step from 2D to 3D in most cases is a minor problem. The major concern in fully 3D computations is the immense increase in the number of unknown variables. 3D time-dependent computations in most cases require an efficient numerical formulation as well as code parallelization. However, computer simulation is one tool for solving the phenomenon of polymer flow and designing the mold geometry such as gate position, cavity thickness, etc. The calculation time is also important for applying and using real production. Many commercial CAE systems have been developed and a 3D FEM system for polymer processing has also been used recently.

A more recent 3D FEM simulation for micro-injection molding and thermal nanoimprinting has been performed which takes the effect of an air trap into account [45]. This 3D simulation of the filling process has been performed using a commercial injection molding CAE system. This CAE system is able to simulate the physical phenomena of various molding processes such as injection press/compression molding, co-injection molding, multilayered injection molding other than normal injection molding, and the function of injection press molding simulation is used for simulation of thermal nanoimprinting in this study. Besides, the system is also able to consider the various calculation functions such as the effect of an air trap, a slip on mold surfaces, surface tension, etc., which are thought to be related to the evaluation of the replication mechanism of micro- and nano-molded surface features.

Generally, the assumption that pressure equals zero ($P = 0$) is imposed on the flow front as a boundary condition in injection molding CAE systems commercially released. However, this assumption does not hold good for the inefficient case of an air vent and for the closed space in the cavity because of the existence of compression resistance of the residual air with progress of the flow front in a cavity. This compression resistance might influence the filling flow behavior in micro- and nanosurface features without any air vent. For the function of an air trap effect in the system [45], compression resistance of the residual air is considered based on the specifications of a set air vent. In the simulations, the Cross–Arrhenius viscous model was used to consider the temperature dependency of shear-thinning fluid and the Spencer–Gilmore equation was used to express PVT behavior. Temperature dependencies of thermal conductivity and specific heat were also considered in the simulation.

Figure 3.15 shows the shape of the molded product used for micro-injection molding. The product consists of two pieces of plate with various micro-surface features, respectively. One is a grid of 400 μm squares of 100 μm height. Each grid is located in every 100 μm space. Another feature consists of four groups of lines and spaces, aligned in the melt flow direction. Lines in each group have different widths and depths, as shown in the figure. The cavity thickness is 0.3 mm in both plates. The solid mesh model consisting of hexahedron elements used in injection molding simulation is shown in Figure 3.16.

Figure 3.17 illustrates the calculated flow front pattern in the gate side part of the grid-type cavity. The air traps are seen conspicuously on the upstream side of the grid features of the transversal direction (TD) that is perpendicular to the melt flow direction (MD), but in contrast no air trap

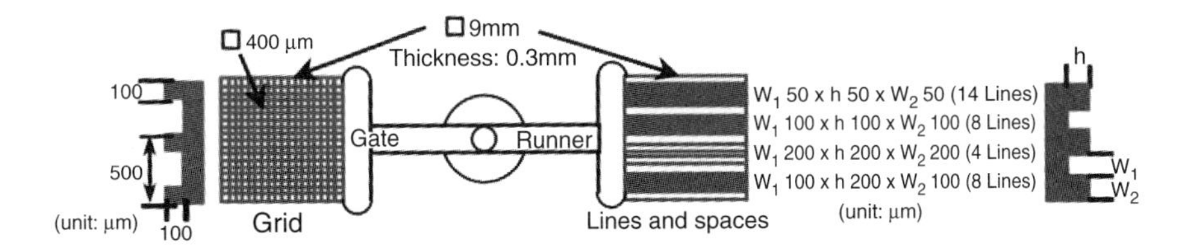

Figure 3.15 Schematic diagrams of molded product with micro-surface features of a grid and lines and spaces

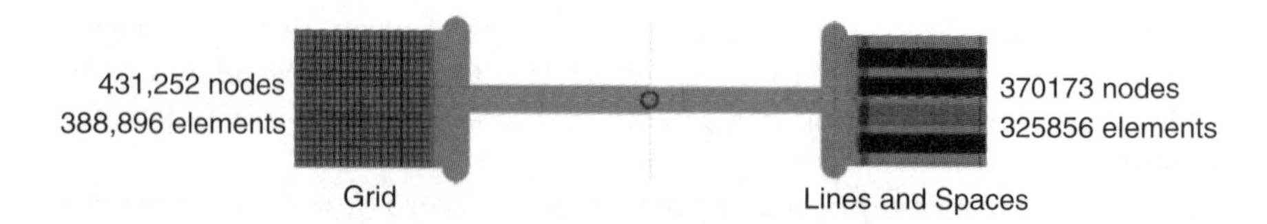

Figure 3.16 Mesh model of a molded product with micro-surface features of a grid and lines and spaces

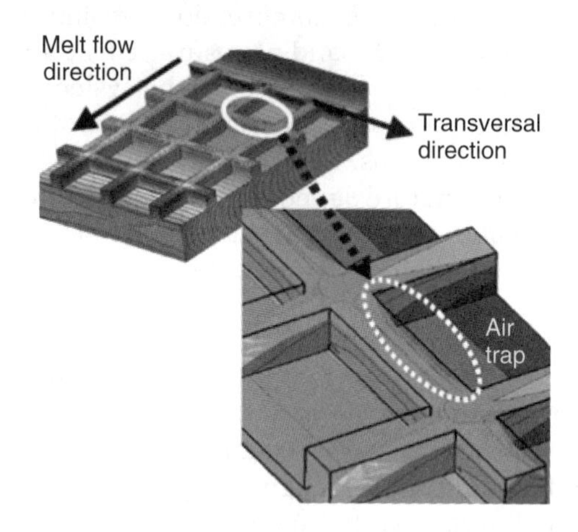

Figure 3.17 Calculated flow front pattern and air traps of the grid-type cavity

exists on the downstream side of the grid features of the TD and in the grid features of the MD. Experimental results also show that a trace like an air trap is observed only on the upstream side of the grid features of the TD, but such a trace is not observed on the downstream side of the grid features of the TD and in the grid features of the MD, and consequently the replication rate of the grid features of the TD is lower whereas that of the MD is higher. These calculations and experimental results suggest that the replication shape and the replication rate are caused by the filling flow behavior in micro-surface features. Figure 3.18 shows the calculated flow front pattern on the gate-side part of the lines and spaces type cavity. It is seen that air is shut in at the corners of the line features. For a cause of these hollows seen in a real molded product, an air trap in the filling stage and/or shrinkage in the cooling stage and/or deformation at mold release are considered. However, because the location and shape of these hollows in a molded product correspond remarkably to those of the air traps

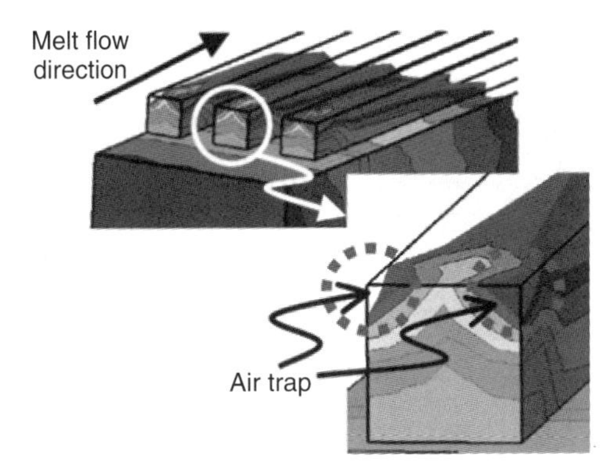

Figure 3.18 Calculated flow front pattern and air traps of the lines and spaces-type cavity

appearing in the calculated flow front pattern, it is concluded that an air trap generated by the filling flow is the main cause of these hollows. Therefore, it is also suggested that the filling flow behavior is very important in evaluating replication shape and replication ratio.

Tada *et al.* [45] focused on the flow behavior for nanosurface features in the filling stage of thermal nanoimprinting. They performed 3D FEM simulations, which take the effect of an air trap into account. The simulation results were also discussed in comparison with experimental results and revealed these effects on the replication mechanism.

The 3D simulation of the filling process in thermal nanoimprinting has been performed using a commercial injection molding CAE system, and is able to consider the various calculation functions such as the effect of an air trap, a slip on mold surfaces, surface tension, etc., which are thought to be related to evaluation of the replication mechanism of nano-molded surface features. In the model, the function of an air trap effect was considered in the flow simulation for this nano molding. Generally, the assumption that pressure equals zero ($P = 0$) is imposed on the flow front as a boundary condition in commercially released injection molding CAE systems. However, this assumption does not hold good for the inefficient case of an air vent and for the closed space in the cavity, as described earlier.

The mold is the same honeycomb structure with depth of 330 nm, width of 8.45 μm, and wall thickness of 1.2 μm (see Figure 3.12). The aspect ratio (depth/width) of nanosurface features is considerably small. The solid mesh model consisting of hexahedron and prism elements used in thermal imprinting simulations is shown in Figure 3.19.

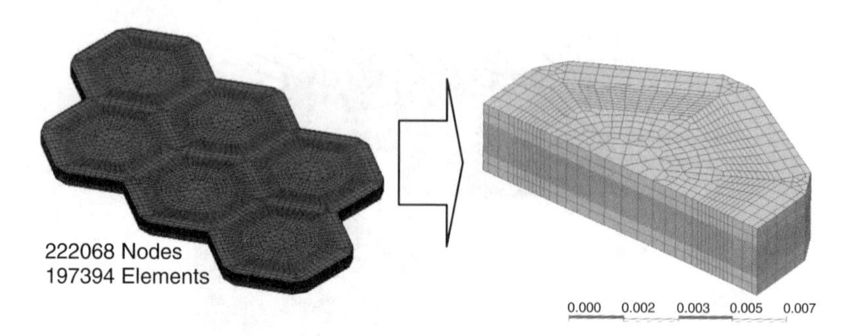

Figure 3.19 Mesh model for thermal nanoimprinting

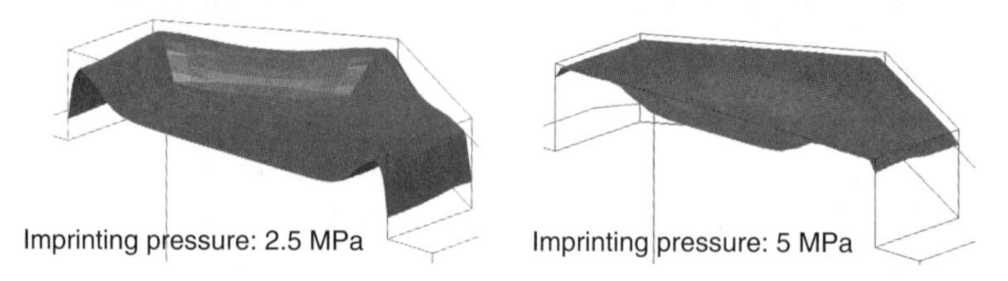

Figure 3.20 Calculated flow front pattern in thermal nanoimprinting (temperature 165 °C)

Figure 3.20 illustrates the calculated flow front pattern under imprinting pressure of 2.5 and 5 MPa in the thermal imprinting simulation. Because the aspect ratio of nanosurface features is so small, penetration of the polymer melt near the wall of the circumference goes ahead from the central part of the cavity. It has been confirmed that penetration flow of the central part of the cavity goes ahead in the surface features with large aspect ratio by other numerical simulations; therefore it is considered that the aspect ratio has a large effect on the penetration behavior. In comparison with the flow front behavior under both imprinting pressure conditions, penetration is faster and the replication rate is higher with respect to the increase in imprinting pressure and it may be said that the imprinting pressure influences the replication rate greatly.

References

[1] Chen, C.-C.A., Chang, S.-W., and Weng, C.-J. 2008. Design and fabrication of optical homogenizer with micro structure by injection molding process. *Proc. SPIE*, Vol. 7058.

[2] Lu, Z. and Zhang, K.F. 2009. Morphology and mechanical properties of polypropylene micro-arrays by micro-injection molding. *Int. J. Adv. Manuf. Technol.* **40**: 490–496.
[3] Katoh, T., Tokuno, R., Zhang, Y., Abe, M., Akita, K., and Akamatsu, M. 2008. Micro injection molding for mass production using LIGA mold inserts. *Microsyst. Technol.* **14**: 1507–1514.
[4] Park, K., Kim, B., and Yao, D. 2006. Numerical simulation for injection molding with a rapidly heated mold, Part I: Flow simulation for thin wall parts. *Polym.-Plast. Technol. Eng.* **45**: 897–902.
[5] Park, K., Kim, B., and Yao, D. 2006. Numerical simulation for injection molding with a rapidly heated mold, Part II: Birefringence prediction. *Polym.-Plast. Technol. Eng.* **45**: 903–909.
[6] Huang, M.-S. and Tai, N.-S. 2009. Experimental rapid surface heating by induction for micro-injection molding of light-guided plates. *J. Appl. Polym. Sci.* **113**: 1345–1354.
[7] Yu, M.-C., Young, W.-B., and Hsu, P.-M. 2007. Micro-injection molding with the infrared assisted mold heating system. *Mater. Sci. Eng. A* **460/461**: 288–295.
[8] Chou, S. Y., Krauss, P. R., and Renstrom, P. J. 1995. Imprint of sub-25 nm vias and trehches in polymers. *Appl. Phys. Lett.* **67**: 3114.
[9] Youn, S.W., Goto, H., Takahashi, M., Ogiwara, M., and Maeda, R. 2007. Thermal imprint process of parylene for MEMS applications. *Key Eng. Mater.* **340/341**(II): 931–936.
[10] Heyderman, L.J., Schifta, H., Davida, C., Gobrechta, J., and Schweizerb, T. 2000. Flow behavior of thin polymer films used for hot embossing lithography. *Microelectron. Eng.* **54**: 229–245.
[11] Eriksson, T. and Rasmussen, H.K. 2005. The effects of polymer melt rheology on the replication of surface microstructures in isothermal molding. *J. Non-Newton. Fluid Mech.* **127**: 191–200.
[12] Juang, Y.-J., Lee, L.J., and Koelling, K.W. 2002. Hot embossing in microfabrication. Part I: Experimental. *Polym. Eng. Sci.* **42**: 539–550.
[13] Juang, Y.-J., Lee, L.J., and Koelling, K.W. 2002. Hot embossing in microfabrication. Part II: Rheological characterization and process analysis. *Polym. Eng. Sci.* **42**: 551–566.
[14] Yao, D., Virupaksha, V. L., and Kim, B. 2005. Study on squeezing flow during nonisothermal embossing of polymer microstructures. *Polym. Eng. Sci.* **45**: 652–660.
[15] Takagi, H., Takahashi, M., Maeda, R., Onishi, Y., Iriye, Y., Iwasaki, T., and Hirai, Y. 2008. Analysis of time dependent polymer deformation based on viscoelastic model in thermal imprint process. *Microelectron. Eng.* **85**: 902–906.
[16] Onishi, Y., Hirai, Y., Takagi, H., Takahashi, M., Tanabe, T., Maeda, R., and Iriye, Y. 2008. Numerical and experimental analysis of intermittent line-and-space patterns in thermal nanoimprint. *Jpn. J. Appl. Phys.* **47**: 5145–5150.
[17] Chou, S.Y., Krauss, P.R. and Renstorm, P.J. 1996. *J. Vac. Sci. Technol. B* **14**: 4129.
[18] Guo, L.J. 2005. *J. Phys. D* **37**: R123.

[19] Hardt, D., Ganesan, B., Qi, W., Dirckx, M. and Rzepniewski, A. Innovation in Manufacturing Systems and Technology (IMST). http://hdl.handle.net/1721.1/3917
[20] Heidari, B., Maximov, I., and Montelius, L. 2000. *J. Vac. Sci. Technol. B* **18**: 3557.
[21] Rowland, H.D. and King, W.P. 2004. *J. Micromech. Microeng.* **14**: 1625.
[22] Shen, X.-J., Pan, L.-W., and Lin, L. 2002. *Sens. Act. A* **97/98**: 428.
[23] Sunder, V.C., Eisler, H.-J., Deng, T., Chan, Y., Thomas, E.L. and Bawendi, M.G. 2004. *Adv. Mater.* **16**: 2137.
[24] Tan, H., Gilbertson, A. and Chou, S.Y. 1998. *J. Vac. Sci. Technol. B* **16**: 3926.
[25] Torres, C.M.S. 2003. *Mater. Sci. Eng. C* **23**: 23.
[26] Youn, S.-W., Goto, H., Takahashi, M., Oyama, S., Oshinomi, Y., Matsutani, K., and Maeda, R. 2007. *J. Micromech. Microeng.* **17**: 1.
[27] Hirai, Y., Fujiwara, M., Okuno, T., Tanaka, Y., Endo, M., Irie, S. *et al.* 2001. *J. Vac. Sci. Technol. B* **19**: 2811.
[28] Hirai, Y., Konishi, T., Yoshikawa, T., and Yoshida, S. 2004. *J. Vac. Sci. Technol. B* **22**: 3288.
[29] Chaix, N., Gourgon, C., Landis, S., Perret, C., Fink, M., Reuther, F., and Mecerreyes, D. 2006. *Nanotechnol.* **17**: 4082.
[30] Landisa, S., Chaixb, N., Hermelina, D., Levedera, T., and Gourgon, C. 2007. *Microelectron. Eng.* **84**: 940.
[31] Schulz, H., Wissen, M., Bogdanski, N., Scheer, H.C., Mattes, K., and Friedrich, Ch. 2006. *Microelectron. Eng.* **83**: 259.
[32] Singh, P., Radhakrishnan, V., and Narayan, K.A. 1990. *Ingenieur-Archiv* **60**: 274.
[33] Schift, H. and Heyderman, L.J. 2003. *Alternative Lithography: Unleashing the Potential of Nanotechnology*. New York: Kluwer Academic, chapter 4, p. 46.
[34] Schift, H., Bellini, S., Gobrecht, J., Reuther, F., Kubenz, M., Mikkelsen, M.B., and Vogelsang, K. 2007. *Microelectron. Eng.* **84**: 932.
[35] Chen, B.S. and Liu, W.H. 1989. *Polym. Eng. Sci.* **29**: 1039–1050.
[36] Huamin, Z. and Dequn, L. *Polym.-Plast. Technol. Eng.* **41**: 91–1021.
[37] Hieber, C.A. and Shen, S.F. 1980. *J. Non-Newton. Fluid Mech.* **7**: 1–32.
[38] Inoue, Y., Higashi, T., and Matsuoka, T. 1996. ANTEC tech. paper, pp. 744–748.
[39] Hwang, C.J. and Kwon, T.H. 2002. *Polym. Eng. Sci.* **42**: 33–50.
[40] Hieber, C.A. 1987. *Injection and Compression Molding Fundamentals*. New York: Marcel Dekker.
[41] Geng, T., Li, D., and Zhou, H. 2006. *Eng. Comput.* **21**: 289–295.
[42] Peters, E.A.J.F., Hulsen, M.A., and van den Brule, B.H.A.A. 2000. *J. Non-Newton. Fluid Mech.* **89**: 209–228.
[43] Wapperom, P. and Keunings, R. 2001. *J. Non-Newton. Fluid Mech.* **95**: 67–83.
[44] Rasmussen, H.K. 2000. *J. Non-Newton. Fluid Mech.* **92**: 227–243.
[45] Tada, K., Fukuzawa, D., Watanabe, A., and Ito, H. 2010. *Rubber Compos.* **39**: 321–326.

4

Mold Fabrication Process

Mitsunori Kokubo[a], Gaku Suzuki[b], and Masao Otaki[b]
[a]*Toshiba Machine Co., Ltd., Japan*
[b]*Toppan Printing Co., Ltd., Japan*

4.1 Ultra Precision Cutting Techniques Applied to Metal Molds Fabrication for Nanoimprint Lithography

4.1.1 Introduction

Molds for nanoimprint lithography with a below micrometer size pattern are usually made using photolithographic techniques of mask exposure and direct electron beam drawing, etc. The method of forming a fine shape to roll for nanoimprint use with the general "roll-to-roll" (RtR) process is shown below. A thin-shaped metallic mold is reproduced from the master mold manufactured using photolithographic techniques with electroformed technology, and is wrapped around the surface of the roll. In contrast, cutting (machining) is suggested as a suitable method to give fine shape directly to the roll for RtR nanoimprinting. Cutting is no more accurate than the photolithographic technique, but the metal can be processed directly without manufacturing the electroformed mold. Hence, it is a method by

Nanoimprint Technology: Nanotransfer for Thermoplastic and Photocurable Polymers, First Edition.
Edited by Jun Taniguchi, Hiroshi Ito, Jun Mizuno, and Takushi Saito.

which a fine-shaped pattern can be formed directly on a metallic roll for RtR nanoimprinting. The RtR process is a nanoimprinting method suitable for adjusting to a large area, can enhance productivity, and is a very effective method for producing various optical films used with FPD such as liquid crystal displays. Recently, the demand for optical sheets is a direction in which it has expanded and proved efficient, for reasons such as its ability to spread and enlarge the liquid crystal display, and satisfy the demand for high brightness, etc. In this way, it has led to advances in the fineness of shape possible on the surface of optical sheets. And the expectations for RtR nanoimprinting as a manufacturing method are great, where a highly accurate transcript is possible. The expectations with such a background using cutting have risen, too. Further, in this chapter we introduce ultra precise machining technology for fine shape using a diamond tool.

4.1.2 Cutting of Fine Groove Shape

A rapid method of highly accurate cutting has been developed, starting with ultra precision processing machines equipped with air bearing spindle. Metal mold processing for a plastic injection machine with aspherical lens became possible through the table feeding and positioning performance of the processing machine and the improvement in monocrystalline diamond tool manufacture technology in the 1980s. Work on fine shape processing began at about the same time. The cutting technology for a concentric circle fine groove of depth 1 μm or less was developed, and applied to optical parts for an optical pick-up mechanism [1]. Afterwards, metal mold processing of an aspherical lens to give a saw diffraction groove became possible, too. The range of application of the (then) super-precision cutting technology was limited mostly to flat or curved surface shapes of rotation symmetry. However, a four-axis control processing machine with one nanometer control was developed after 1990, and the mirror surface finish of a free and curved surface shape became possible. In addition, the fine groove processing technology of a diffraction optics element mold with free curve surface was developed. Because the means to mass produce an efficient optical element cheaply had been limited to the injection molding method, the processing object stayed comparatively small. However, it can be said that a basic technology for fine groove formation by cutting had already been established. Recently, a light-guiding plate with fine groove pattern structure came to be manufactured, owing to the demand for high brightness of LCDs. For the enlargement of the display to advance at the same time, and to correspond to the product development, large-scale, ultra-precision processing machines came to be developed one after another. Enlargement was advanced as with the groove processing machine for the mold of a light-guiding plate, and highly accurate cutting of a fine groove became possible.

4.1.3 Method of Cutting Groove

In general, there are two methods of processing a fine groove by cutting with high accuracy. First of all, there is a turning/planing cutting method processed with the cutting tool fixed to a table. Then, there is a fly cutting method which rotates the cutting tool. Figure 4.1 outlines the turning/planing cutting method and the fly cutting method. Both methods transcribe the shape of the diamond tool blade tip in forming the groove. The turning/planing cutting method is a "pull cutting" process, in which the table feed speed of the processing machine is assumed to be the cutting speed. In contrast, the fly cutting method processes the groove while rotating the tool with a static pressure air bearing spindle, etc., and the surrounding speed of tool rotation is the "cutting speed." Setting the table feed speed of the machine according to the speed of tool rotation is necessary, and the processing efficiency is lower for than the planing cutting method. However, the amount of removal by tool rotation can be small, and a high cutting speed is obtained. In addition, the generation of burr in the ridge line when a matrix shape pattern is formed by the method of intersection of the groove can be controlled.

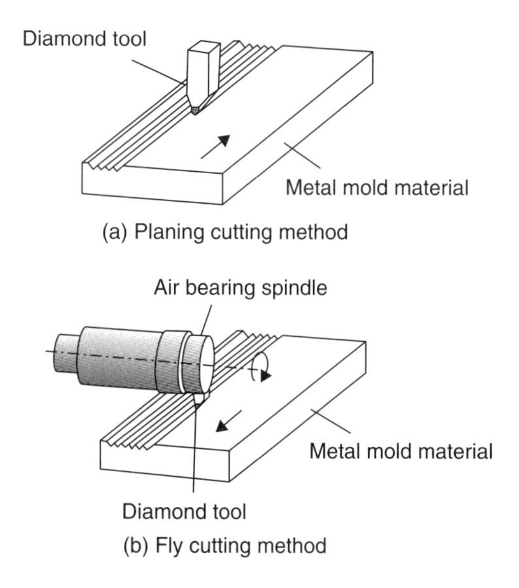

Figure 4.1 Processing method for fine groove

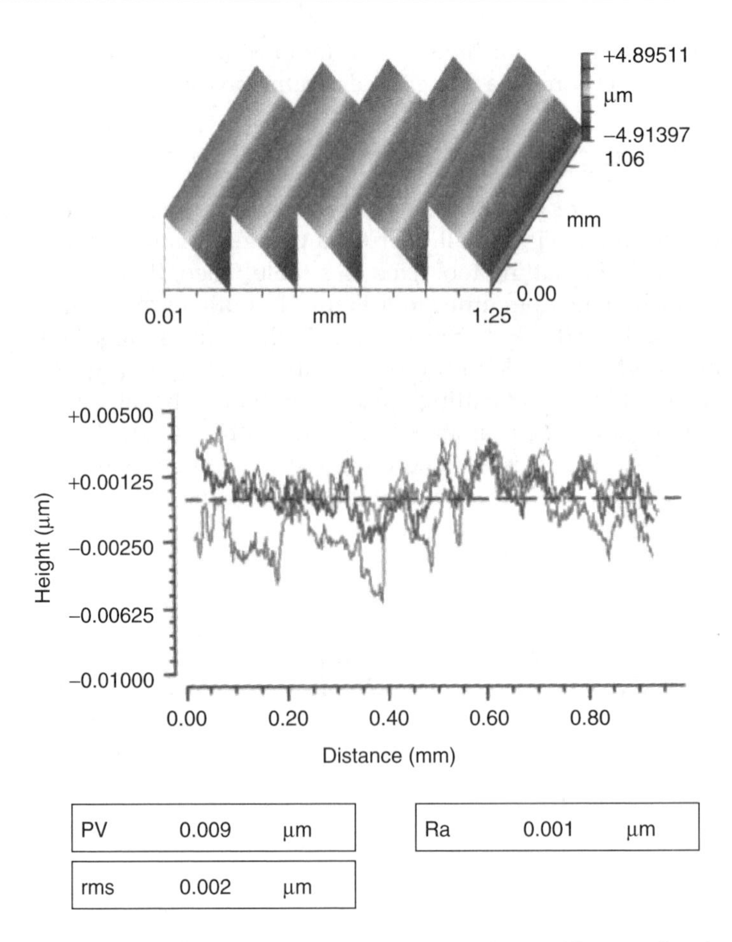

Figure 4.2 Surface roughness by the planing cutting method

Figure 4.2 shows the processing surface accuracy measurement result (measurement machine: New View, ZYGO) of "saw grooves" (material: electroless Ni plating) cut with an ultra precision groove processing machine (UVM-100A, Toshiba Machine Co., Ltd). The groove surface has 1 nm Ra roughness in the direction of cutting, and highly accurate flatness is obtained by the planing cutting method. The pattern in which square pillars and pyramids are regularly arranged can be formed by processing a groove with two or more directions intersecting by the planing cutting method and the fly cutting method.

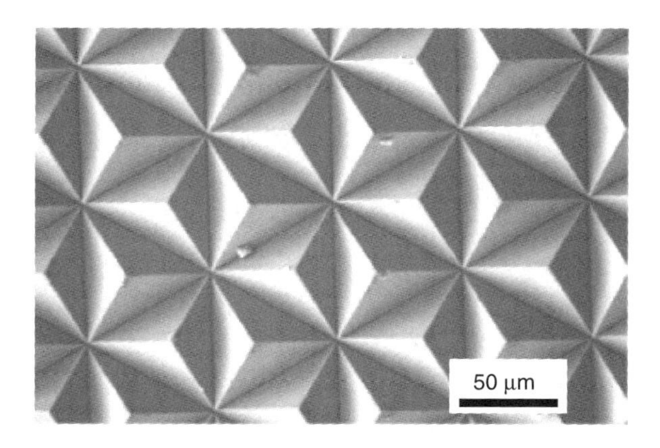

Figure 4.3 Triangular pyramid shape processed by the fly cutting method

Figure 4.3 shows a SEM image of the triangular pyramid shape processed by the fly cutting method. That is, processing of a V-form groove in three directions with an angle of 120° respectively at 50 μm pitch. A triangular pyramid fine shape can be formed with high accuracy.

4.1.4 Precision Cutting of Cylindrical Material

It is possible to cut with precision for a metallic roll by the pyramid shape cutting method described here. A high-precision grooving lathe machine (ULR-628B(H), Toshiba Machine) was used for this processing. Figure 4.4 shows a general view of the ultra precision grooving lathe machine. An oil static pressure bearing is adopted in the spindle of this machine, and it is

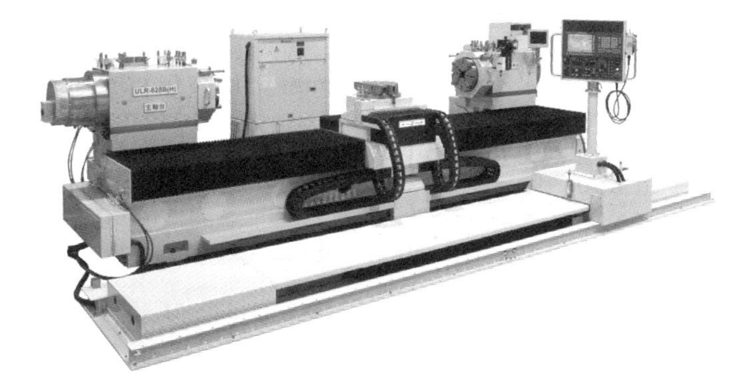

Figure 4.4 Ultra precision grooving lathe machine

Figure 4.5 V-groove processing of the roll

equipped with a numerical control index function. The tool drive stage is driven by a linear motor. Moreover, it is also possible to equip the machine with a static pressure air bearing spindle for fly cutting. Figure 4.5 shows the appearance of V-groove processing of the roll. The diameter of the roll is 250 mm, and the material on the surface is copper plating.

Figure 4.6(a) shows the square pyramid pattern formed by the intersection of an orthogonal V groove. The triangular pyramid pattern formed by the intersection of a V groove in three directions is shown in Figure 4.6(b). Here, each pyramid pattern was formed by the intersection of a 50 μm pitch V groove. The turning method was applied for groove processing in the direction of the roll surroundings and the fly cutting method was applied for processing in the direction of a roll center axis in the processing of the square pyramid pattern. Both grooves were processed using the main axis calculation index function. In contrast, the turning method was applied to all V grooves in three directions when processing the triangular pyramid pattern. The groove was processed with a diagonal "multi article screw cutting" method to control the rotation angle of spindle and tool feeding.

4.1.5 High-speed, Ultra Precision Machining System

Recently, equipment has been developed with a fine drive stage which can operate at high speed in the ultra precision processing machine, moving the

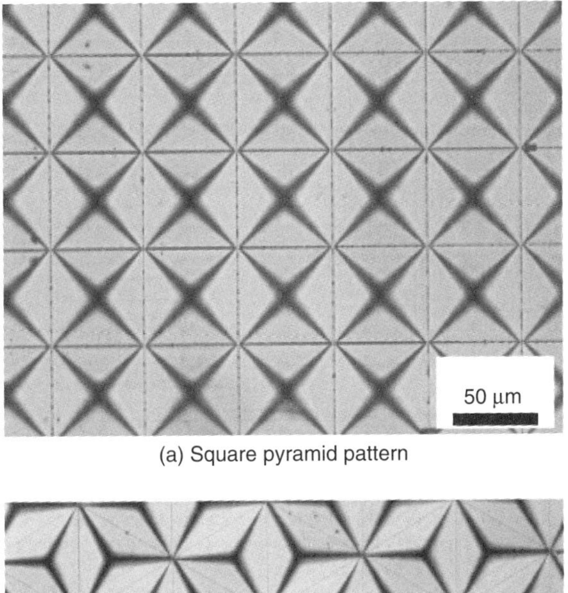

(a) Square pyramid pattern

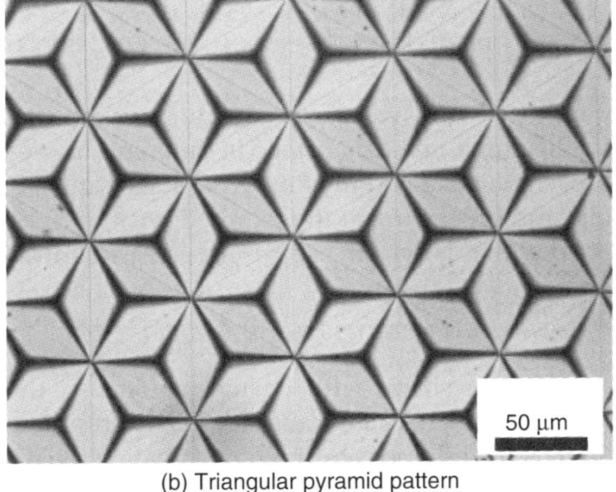

(b) Triangular pyramid pattern

Figure 4.6 Pyramid pattern processed on roll

diamond tool at high speed and processing work pieces. The development of this technique began as a technology to make amends for the movement accuracy of the processing machine. However, it evolved to form a fine three-dimensional shape at high speed. Work has been carried out on a fine processing case using this technology, as recorded in this text for reasons of showing a system technology which limits the purpose of speeding up a fine processing to a "high-speed, ultra precision cutting system." The composition of the system is shown in Figure 4.7. An accumulating PZT actuator is used

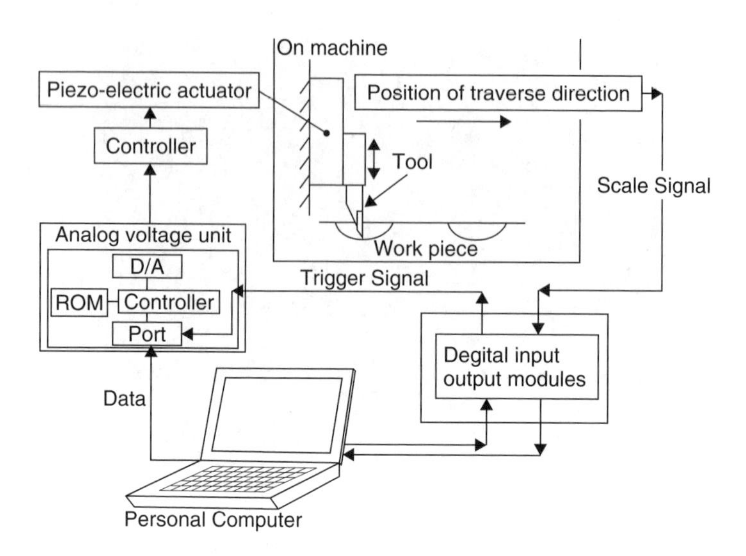

Figure 4.7 Composition of high-speed, fine processing system

for the driving mechanism of this stage. The numerical control system of a processing machine and a fine drive stage is different. However, control of the processing machine table and the fine drive stage can be matched to the operation timing, and processing at the required position enabled. Two fine patterns processed with this system are introduced as follows.

4.1.6 Fine Pattern Processing by Bit Map Data

First, a two-dimensional shape with change pattern in the direction of depth is converted into gray-scale bit map data with resolution of 256 bits. Next, a three-dimensional shape is formed by repeating the planing/cutting method with the scanning lines method. Figure 4.8 shows an example of processing with the high-precision asphric machining process machine (ULG-100D(SH3), Toshiba Machine). The machine was equipped with a fine drive stage. The high-speed, fine processing system was composed, and processing carried out. The cutting operation of the fine drive stage was controlled according to bit map data, the machine table was fed at a speed of 1 m/min, and planing/cutting was carried out.

4.1.7 Machining of Dot Pattern Array

The shuttling operation is done at high speed for a diamond tool on the fine drive stage in the incision direction. The dot pattern for a constant cycle is

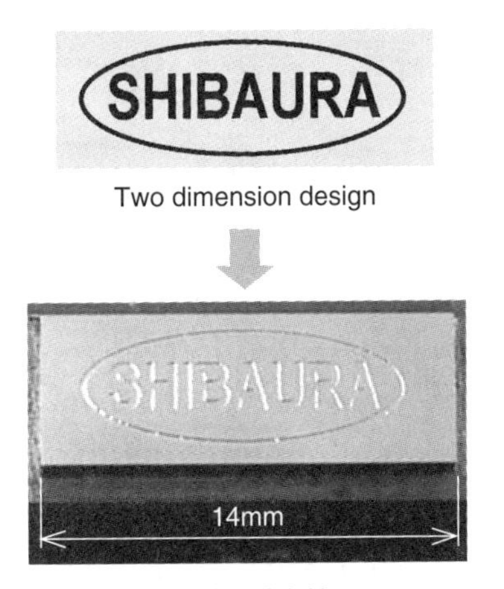

Figure 4.8 Fine pattern processing by bit map data

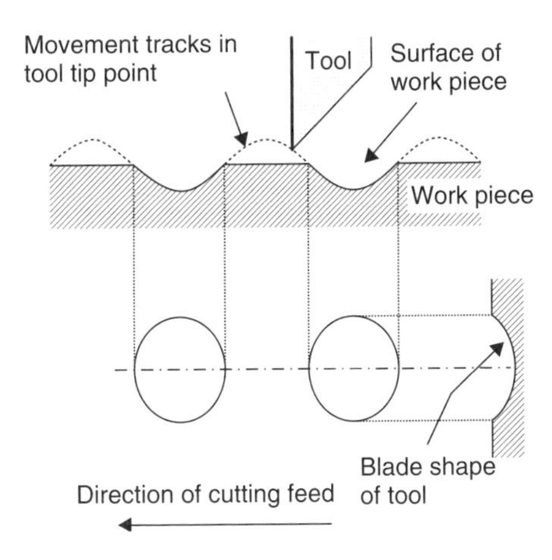

Figure 4.9 Outline of dot pattern array processing

formed by horizontal feeding of the processing machine table, and repeating the intermission cutting at the same time. Figure 4.9 shows the outline of this processing method. The section shape of the dot is decided by the track shape corresponding to the stage drive voltage crimp and the processing machine table feeding speed in the direction of tool feeding. In addition, the

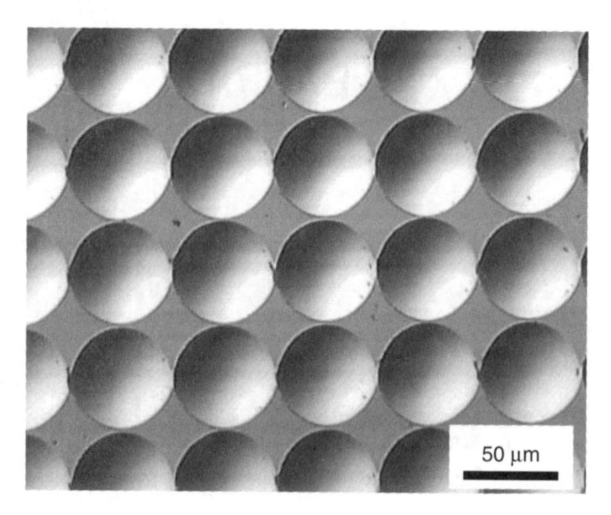

Figure 4.10 Round dot pattern processed on plane substrate with tool feeding

section shape of the dot is decided by the tool blade shape in an orthogonal direction to the direction of tool feeding. For instance, consider the case of a tool with a circular arc blade. In that case, the curved surface shape of the dot becomes a deformation cylinder shape by which the movement tracks of the tool are assumed to follow a curved axis.

Figure 4.10 is an example of processing a dot pattern. The high-precision asphric machining process machine ULG-100D(SH3) was used as processing device. The pattern, in which round dots are adjacent, was processed on a substrate with copper plating given. The dots have diameter 53.5 μm, 2 μm depth, and the processing time required to put one dot in place is 13 ms. The round dot shape is processed with high accuracy.

An example applying the dot pattern processing by this method to the surface of a metallic roll is introduced. Figure 4.11 shows the processing with a high-precision grooving lathe machine (ULR-628B(H), Toshiba Machine). The machine is equipped with a high-speed, fine processing system just as in Figure 4.8. The diameter of the roll is 200 mm, and the material on the surface is electroless Ni–P plating. Figure 4.12 shows a photograph of the dot pattern processed on the surface of the roll. For a dot depth of 5 μm, the processing time required to place one dot is 1.1 ms. The high-speed, fine processing system was able to process a fine dot on the surface of the metallic roll regularly.

Figure 4.11 Dot pattern processing on the surface of a metallic roll

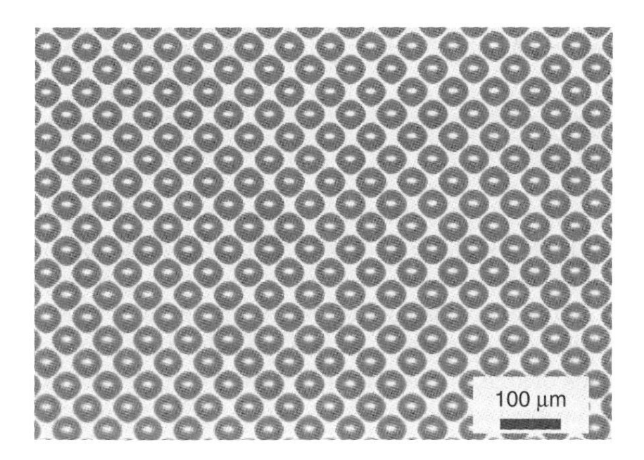

Figure 4.12 Photograph of the dot pattern processed on the surface of the roll. Image courtesy of ELIONIX

4.1.8 Improvement Points of the System

Here we describe the points for improvement that should be considered in future systems. Firstly, there is a processing shape error which originates in the dynamic characteristic of the movement object and occurs when the dot shape, etc. is processed by the system. The reason for this is that a number of problems occur when the movement, including high cycle movement, is carried out: a decrease in real amplitude against the instruction amplitude, a phase delay of the movement crimp, etc. Therefore, it is necessary to improve the stage dynamic characteristics and the techniques of error correction. Moreover, it is recorded that the range of external dot sizes and depths that can be processed is restricted by the tip of the blade shape – such as the clearance angle of the diamond tool used for processing.

4.1.9 Summary

In this section, the ultra precision cutting technology of a minute shape with a diamond tool was introduced. We forecast that the demand for making fine and high-accuracy shapes will increase, and a further improvement in processing efficiency will be needed in the future. It is thought that the necessity for additional technologies like high-speed, fine processing systems will increase to improve the accuracy and speed of processing machines.

4.2 Nanoimprint Mold Fabrication Using Photomask Technology

4.2.1 Introduction

Since its early stage of development, NIL has shown sub-100 nm printability features [2]. At an early stage of research, it was considered good enough if the required area of mold pattern was very small, for example, less than 100 μm square. However, larger molds with an area of over several dozen square millimeters are required in practice. In addition, extremely high precision in terms of size, shape, and placement is required for patterns from these molds. It has been critical for the commercialization of NIL to obtain a mold with high precision and high resolution at a reasonable cost.

In responding to the requirements, it is reasonable to apply photomask fabrication technology, which has been used in photolithography [3, 4]. For photomask fabrication, EB lithography and dry etching are important processes to form fine patterns. EB lithography is a patterning process on a resist layer by EB writer. The actual patterns are obtained from the pattern data designed by a CAD (computer-aided design) system. Dry etching is used for pattern transfer to the chrome (Cr) film on the quartz (QZ) substrate from the resist pattern. In addition to Cr film etching, etching of the QZ substrate is required for use as a nanoimprint mold.

Mold fabrication processes using the photomask fabrication technique are discussed here, especially focusing on the EB lithography and dry etching technologies. Actual examples of mold structures manufactured by these technologies will be discussed.

4.2.2 Summary of Mold Manufacturing Process

4.2.2.1 QZ Mold for UV Imprint

The standard photomask process requires etching only a Cr thin-film deposit on a QZ substrate, while the advanced photomask process requires etching some parts of a QZ substrate in addition to the Cr film etching. This advanced photomask technology is applied for QZ mold fabrication, moreover, further development is added to fabricate even finer patterns.

Figure 4.13 indicates the standard process flow for QZ mold fabrication. At first, Cr film is sputtered onto one side of a 6025 (6 inch square by 0.25 inch thickness) QZ substrate. This Cr film carries out not only the function of a hard mask for QZ dry etching, but also the electrification prevention film for EB writing. When the EB resist is spin-coated onto the QZ substrate, its surface is electrified by the EB exposure because QZ is an insulator. Conductive Cr film is needed on the surface to prevent this problem. Generally, the Cr film thickness is less than the general photomask's thickness for the purpose of improving pattern resolution.

Generally, a positive-type EB resist is spin-coated onto the Cr film, and then its substrate is heated with a hot plate. Subsequent to EB drawing, a positive EB resist is developed to resolve the exposed area with an alkaline-type developer. After rinsing and drying treatments, the required patterns are formed on the Cr film. These processes together are called "EB lithography."

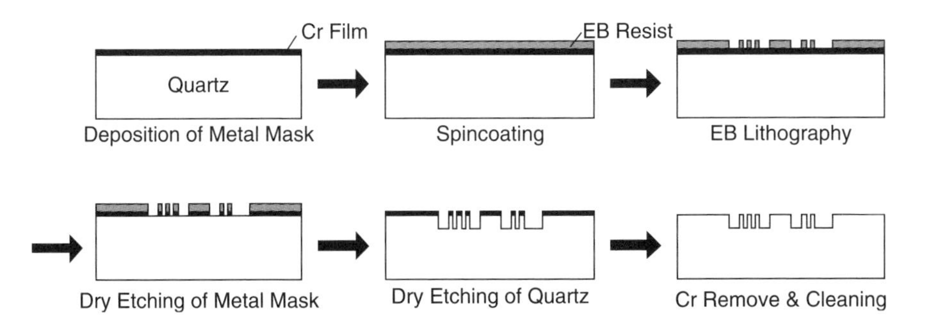

Figure 4.13 Process flow of QZ mold fabrication

However, it is very difficult to form fine patterns (less than 100 nm wide, for example). So, the process conditions must be set carefully.

The EB resist thickness is fixed by considering both the required minimum pattern width and the Cr film thickness. A thinner resist is advantageous to obtain a smaller pattern for EB lithography. The adhesive force between the resist and the Cr film surface becomes weaker as the pattern becomes smaller, and that makes the process susceptible to resist patterns peeling off or collapsing with pressure at the development or rinse processes or by surface tension at the drying process. If the height of the resist pattern is low, those external influences could be reduced, which would support the formation of smaller patterns. However, since the resist pattern is used as an etch mask for the following Cr film etching, an adequate resist thickness is required. As the Cr film becomes thicker, a thicker resist layer is needed accordingly, which makes the formation of fine resist patterns difficult. Thus, the Cr film needs to be as thin as possible to etch the QZ substrate sufficiently.

Up to this point, the process is the same as for binary-type photomask fabrication. For mold fabrication, etching is extended to the surface of the QZ substrate using the etched Cr pattern as a hard mask.

Finally, wet cleaning is performed to remove the Cr film and dry-etching residue on the substrate, and then we obtain a QZ mold with our designed features formed on its surface. This mold pattern will be evaluated with an inspection tool, such as CD-SEM or AFM. Unfortunately, a feature width less than 100 nm is difficult to inspect rapidly and accurately, as no effective method has been established yet. Moreover, even if a defect is detected, we do not have any repair technology for nanoimprint molds (although such technology is already established in the photomask field). Mold patterns used for storage media or optics allow defects up to some level, but those for semiconductors need to be repaired. It is necessary to establish defect inspection and repair techniques for nanoimprint molds to apply the latest frontiers of semiconductor technology.

4.2.2.2 Silicon Mold for Thermal Imprint

The fabrication technologies of Si stencil masks (EB character projection aperture for DW (direct writing)) are known as nanopattern fabrication with Si substrate. EBDW once attracted much attention and was studied extensively, considering that it could be a breakthrough in terms of photolithography limitation. However, because the EBDW throughput didn't reach that of photolithography, EBDW did not become mainstream. As for photolithography, it will have prolonged life via immersion lithography and double patterning technology. In the development of Si stencil masks, dry-etching technology for fabricating fine trench and hole patterns has been established. Today, this technology could be utilized to fabricate nanoimprint molds.

The fabrication flow of an Si mold is similar to Figure 4.13, oriented by EB lithography and dry etching. However, unlike the QZ substrate, since the Si substrate is not an insulator, it would not charge through EB exposure even without the Cr film. In this way, when an EB resist is used as an etching mask for Si, the fabrication process would be reduced by eliminating Cr film deposition and its dry-etching processes.

An etching technique with high selectivity for the resist mask was generally developed in bulk micromachining. In most cases, deep structures with relatively high A/R (aspect ratio) are fabricated. Thermal nanoimprint technology involves a thermal cycle process at the time of transferring patterns, where the mold or substrate is susceptible to expansion/contraction. This indicates that transferring fine patterns with high position accuracy is difficult. QZ molds used for UV-NIL are more suitable for fine pattern transcription. In contrast, Si molds could be utilized in thermal NIL for their flexible workability, and applied to microfluidics and biotechnology that require molds with high A/R structure.

4.2.2.3 Nickel Mold for Thermal Imprint

A metal mold (stamper) carrying fine, rugged patterns on its surface has been used to manufacture optical memory disks. In most cases, materials for these stampers are Ni and Ni alloys. The manufacturing process of such a stamper is applied for Ni mold fabrication. Unlike photomasks, QZ or Si molds, fine patterns on these Ni molds would not be fabricated through a dry-etching process. The main reason is that since Ni compounds rarely have a high vapor pressure, it is impossible to volatilize the reaction products by dry etching. Although chemical wet etching is possible, this is not suitable for fabricating fine and dense patterns required for nanoimprint molds.

Alternatively, an electroforming process is used to fabricate Ni molds with fine patterns on their surface. Figure 4.14 indicates the process flow for fabricating an Ni mold. The basic flow is shown in Figure 4.14(i). A fine EB resist pattern made by EB lithography is used as a master for the electroplating process. A seed layer required for electroplating is deposited by vapor deposition, sputtering, or electroless Ni plating. After electroplating, Ni is finally peeled off from the resist master. Figure 4.15 shows examples of Ni molds fabricated through EB resist masters.

As shown in Figure 4.14(ii), electroplating could be done using etched Si substrate as an original master pattern. This method of fabricating a high-A/R or 3D structure is difficult using an EB resist master pattern. As indicated in Figure 4.14(iii), a photoresist pattern could be used as a master

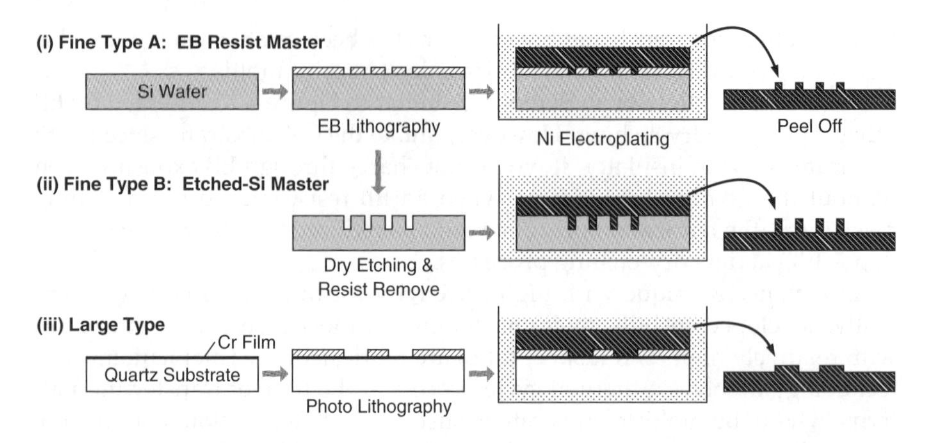

Figure 4.14 Process flow of Ni mold fabrication

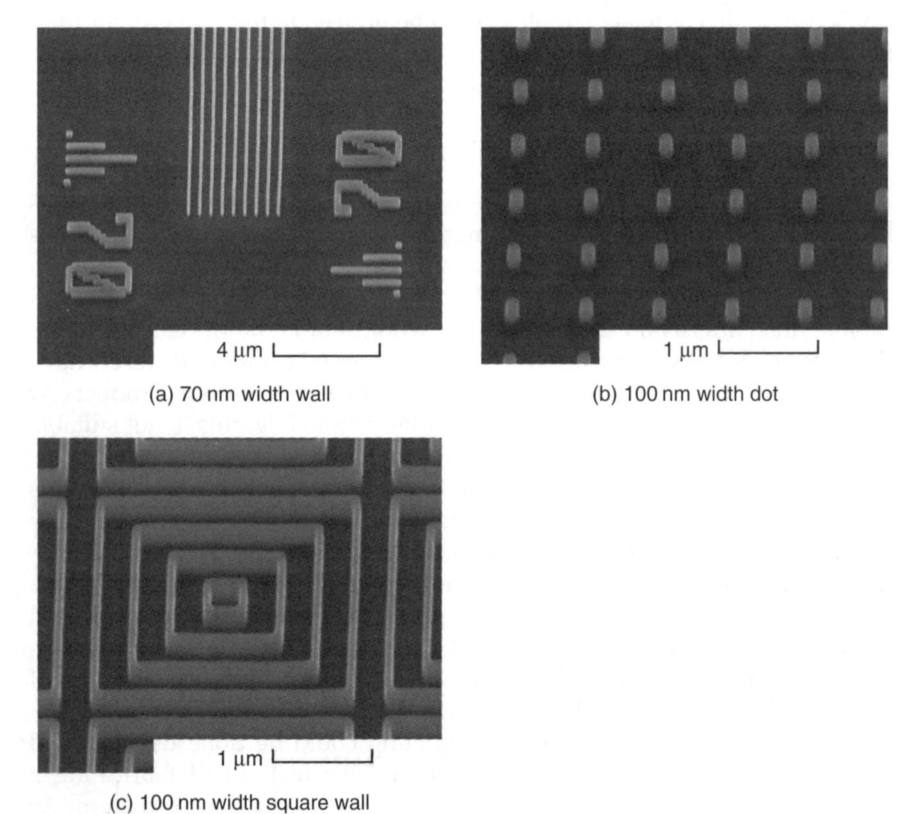

(a) 70 nm width wall

(b) 100 nm width dot

(c) 100 nm width square wall

Figure 4.15 Examples of electroplated Ni structures from EB resist master

pattern. This technique is especially useful for obtaining a larger pattern area than by EB drawing.

Not only the manufacturing processes, but also the mechanical characteristics of Ni (polycrystalline metal) are different from Si (single crystal), or QZ (amorphous). For example, the main damage to mold patterns is brittle destruction in QZ and Si molds whilst it is plastic deformation in Ni molds. Considering such factors, an Ni mold could be considered as similar to a resin mold like PDMS, etc. However, an Ni mold has some characteristics that a resin mold would not carry: fine patterns can easily be replicated from the EB resist or from imprinted resin; it is resistant to temperature and pressure for the thermal nanoimprint; its elastic deformation is much smaller than that of resin. These characteristics are especially suitable for molds used in RtR thermal nanoimprinting. An Ni mold is expected to be used over a wider range to apply nanoimprint technology in various fields.

4.2.3 Pattern Writing Technique

4.2.3.1 Writing Methodology by Pattern Size and Writing Area

Generally, imprint molds (templates) for nanoimprint technology are written by EB or laser beam. Writing technology to fabricate the mold pattern will be discussed.

If a pattern is visible to the human eye (0.1 mm or larger), a number of methods are available, such as the printing technique, computer printout, or hand drawing. 3D patterns can be manufactured through direct cutting, such as the machining method or laser beam patterning (ablation). However, for smaller than 0.1 mm patterns, the applicable technology is limited.

The technology for fabrication of minute patterns has been developed primarily in the semiconductor industry and has been industrialized. The pattern size of a semiconductor integrated circuit has been reduced from several micrometers to several tens of nanometers, an approximately 1/100 reduction over the last three decades.

As of 2010, a semiconductor pattern size of 30–45 nm has been put into production. The semiconductor device is printed on an Si wafer using a lithography exposure tool with highly sophisticated lens (stepper), based on photomasks as a photographic negative. The photomask pattern is printed on an Si wafer, which is coated with photosensitive film (photoresist) reduced 4 or 5 times depending on the stepper.

The pattern size on the photomask is 4 or 5 times larger than the semiconductor pattern on the Si wafer. Photomasks have been manufactured using EB lithography tools since the late 1970s, and a huge quantity of designed patterns can be written precisely and quickly [5, 6].

The photomask is made of 100 or 150 mm^2 glass substrate called blanks, sputtered with 0.1 μm of metal compound film including Cr, etc. and coated with photoresist material sensitive to EB exposure. Generation of the mask pattern is written on glass substrate blanks by an EB located in a vacuum chamber. Cr film is used not only as an opaque material for the mask pattern, but also as a conductive material for electrical discharge.

The patterns on the photomask for the semiconductor are dominated by a combination of rectangles with edges located in the x and y directions. Curved lines and oblique lines could also be created by controlling the pattern layout of minute rectangles. It is possible to write a minimum line width up to around 50 nm using the current EB writers for a photomask. Pattern registration and pattern size fidelity on photomasks for high-end semiconductor devices have improved the pattern accuracy to several nanometers in accordance with the requirements of semiconductor manufacture. The diameter of an EB spot can be focused to several micrometers or several nanometers depending on its target application.

In the case of quantum effect or photonic-based devices, a smaller beam system is preferable instead of a photomask EB system. This beam system makes it possible to fabricate even a 10 nm line width. Since no other tool can generate such a minute pattern, EB writers are essential for NIL. Examples of primary EB writers are listed in Table 4.1. Figure 4.16 shows a typical EB system with a variable-shaped beam as listed in Table 4.1 utilized for photomask manufacture.

It should be pointed out that the writing time to generate fine patterns with a fine focused EB is much longer than the time it takes to control the EB and inject enough electrons into the resist. Since tools offering high accuracy and advanced functionality are extremely expensive (several million to several tens of million dollars), writing a large area of fine patterns comes at a high cost.

In contrast, laser writers have been developed for rough pattern fabrication of more than several micrometers. Plates for printing, negative film for print circuit boards, original plates for microfabrication, large-scale masks for flat panel displays, etc. are typically fabricated by laser writers.

Laser writers are also used in photomask fabrication for semiconductor devices, although finer patterns with higher fidelity are required compared with the examples above. For semiconductor device fabrication, a mask set including dozens of masks is fabricated using both EB and laser writers. Relatively rough patterns applicable for several specific semiconductor processes could be written by a laser exposure system.

Since a laser beam emits light, unlike EB, the beam direction cannot be controlled electrically. However, the laser beam can be controlled mechanically

Table 4.1 Performance of recent EB writers

Equipment	JBX-3050MV	EBM-7000	JBX-9300FS	VB-300	ELS-7000	CABL-9000C
Beam source	LaB6	LaB6	ZrO/W (Schottky)	ZrO/W (TFE)	ZrO/W (TFE)	ZrO/W (TFE)
Acceleration voltage (kV)	50	50	100/50	100/50	100/75/50/25	5–50
Beam shape	Variable-shaped beam	Variable-shaped beam	Spot beam	Spot beam	Spot beam	Spot beam
Beam deflection	Vector scan	Vector scan	Vector scan/step & repeat	Vector scan	Vector scan/raster scan	Vector scan/raster scan
Maximum substrate size (mm)	178	152	300/177	300/177	200	200
Minimum pattern size (nm)	–	–	≤20	8	5	≤10
CD uniformity (nm)	3.5	4	–	–	–	–
Field stitching accuracy (nm)	–	–	20	15	20	20–50
Image placement precision (nm)	7	5	25*	15	20*	20–50*
Main application	Semiconductor photomask	Semiconductor photomask	Wafer direct write/nano devices	Wafer direct write/photomask/nano devices	Wafer direct write/photomask/optoelectronic devices	Wafer direct write/photomask/optoelectronic devices
Vendor	JEOL	NuFlare	JEOL	Vistec	ELIONIX	CRESTEC

*Overlay accuracy.

Figure 4.16 EB exposure system, JBX 3050. Image courtesy of JEOL, Ltd.

with fine mirrors or an acousto-optic modulator (AOM), and can fabricate patterns onto a photosensitive film or photoresist on a substrate.

Light exposure tools including laser writers are able to expose a 2 m×2 m sized substrate because they build patterns in the atmosphere. Equipment with a very large and heavy moving stage holding a large substrate for exposure can be manufactured more easily in an air environment than in vacuum. Selecting an appropriate beam size will significantly reduce the exposure time for a plate.

Table 4.2 gives a list of laser writers, mainly capable of fabricating fine patterns. Particularly, the KrF excimer laser could fabricate up to a 220 nm pattern, using an SLM (spatial light modulator).

Every tool listed in Table 4.2 exposes the laser beam to a photoresist on a glass substrate with chrome sputtered (blanks). In contrast, tools for printed circuit boards cannot generate fine patterns, but can write onto photographic film of large area at high speed using long-wavelength light. This film is also used as a photomask to fabricate electronic materials.

Although laser writers are not capable of fabricating fine patterns including those of nanometer order, they are cost-effective since their price is usually one digit smaller than that of EB writers, and they have a relatively shorter writing time.

As for pattern fabrication tools, the fundamental difference between laser beams and EBs is their capability to generate fine patterns. An EB is capable

Table 4.2 Examples of primary laser writers

Equipment	Sigma-7500-II	Omega-6800HA	LRS 15000-TFT3	DWL 2000	ALTA-4300
Beam source	KrF excimer laser	Krypton ion laser	Krypton ion laser	User-specific laser	Ar ion laser
Wavelength (nm)	248	413	413	363–442	257
Beam shape	Partially coherent SLM imaging	Spot beam	Spot beam	Spot beam	Spot beam
Beam deflection	SLM	AOM	Multibeam deflection by AOM	AOM	AOM & polygon mirror
Maximum substrate size (mm)	152	152	1300×1500	200	152
Pattern size precision (nm)	5.5	15	70	60	9
Image placement precision (nm)	12	25	170	70	15
Main application	Semiconductor photomask	Semiconductor photomask	Photomask for FPD	Photomask & wafers for MEMS or semiconductor	Semiconductor photomask
Vendor	Micronic	Micronic	Micronic	Heidelberg Instruments	Applied Materials

AOM: acousto-optic modulator. FPD: flat panel display. SLM: spatial light modulator.

of generating fine patterns at nanometer level, but it is not possible to write patterns several tens of centimeters square, as this takes a longer time and has a larger cost.

Figure 4.17 indicates the relationship between pattern size and dimension for writing methods. The writing tool with a spot beam as indicated in Table 4.1 is able to generate several nanometer patterns of size 300 mm at most, but such patterns take several weeks or more to finish so are unrealistic in practical terms.

If an appropriate writing tool is unavailable, repeat printing is a possible option. Patterns built into a given small area called a chip or die are printed repeatedly to cover the large targeted area. However, it is technically difficult to carry out this process at low cost due to the stitching accuracy (pattern placement accuracy). If the tool cost is affordable, exposure tools or a stepper for semiconductors could be one solution.

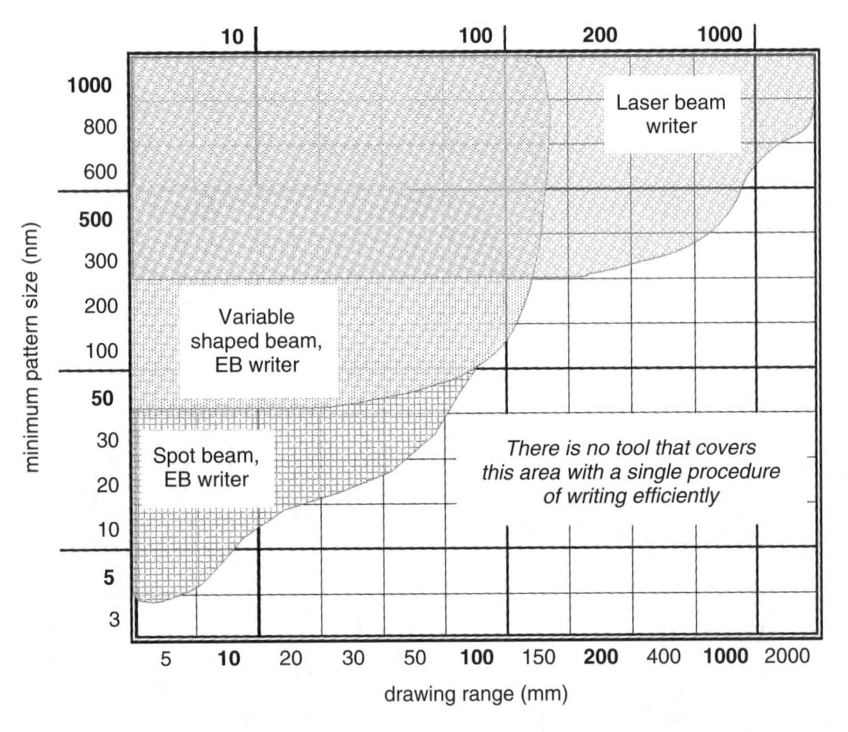

Figure 4.17 Suitable fields depending on mold writing method; relationship between writing area and pattern resolution

4.2.3.2 EB Writing Technique

A laser beam pattern is generated by turning the beam on/off repeatedly, and scanning the entire substrate; the mechanism is similar to the old days of black-and-white TV.

The EB employs a similar technology, but its methods vary depending on the purpose. When its stage is stopped, the EB covers only a limited area (of millimeter order). This area is often called a field. When a larger area needs to be covered, the stage loaded with the substrate will be moved to connect with a neighboring field to generate the entire target pattern.

There are several ways to generate patterns on the field. See Figure 4.18.

Figure 4.18(a) shows a raster scan, the same method as a laser beam scan. In a raster scan, the beam scans the entire field area (even without pattern data). Conversely, a vector scan is a system where the beam scans a selected area only (with pattern data). When the beam shape is "spot beam" in Table 4.1, either method described above is used.

In addition to these, an EB writer called a line-scan mode controlling a thin electron line on the surface is used to build fine patterns including quantum-effect devices. For a spot beam, the beam cross-section is circular with Gaussian distribution.

However, if the pattern is large compared to the beam size, it takes a long time to complete pattern writing even by the method described in Figure 4.18(b).

In response to this issue, a variable-shaped beam technology that could expose a larger pattern in less time has been developed in Japan. Compared to the method for scanning with a thin EB, it improves the writing speed greatly.

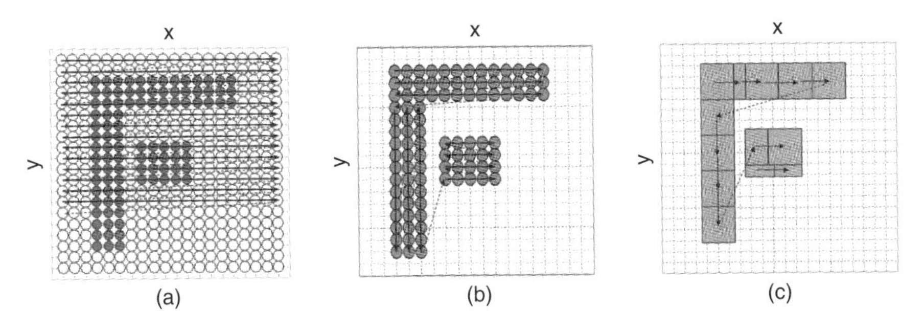

Figure 4.18 (a) Spot beam, raster scan; (b) spot beam, vector scan; (c) variable-shaped beam, vector scan

Taking this method as an example, a control method of EB will be explained as follows. The EB generator which generates the controlled beam is called an EB column (column for short). The writing process will be explained using Figure 4.19. (i) An EB is extracted from the electron source with high voltage (e.g. 50 kV); (ii) it is focused with the first lens; (iii) it is irradiated onto the first aperture top; (iv) it is shaped into a square form as the first variation. (v) The beam location will be shifted with the shaping deflector for a beam passing through the first aperture; the beam axis will be shifted to the second aperture. (vi) The beam is shaped into a rectangle by the second aperture. (vii) The shaped beam is demagnified with a demagnifying lens and deflected by a positioning deflector. (viii) The shaped beam, focused by the projection lens, is exposed to the desired location on the substrate coated with resist. (ix) Patterns to be written onto one field are fractured into fine rectangular form, and each rectangle is shot until the desired pattern

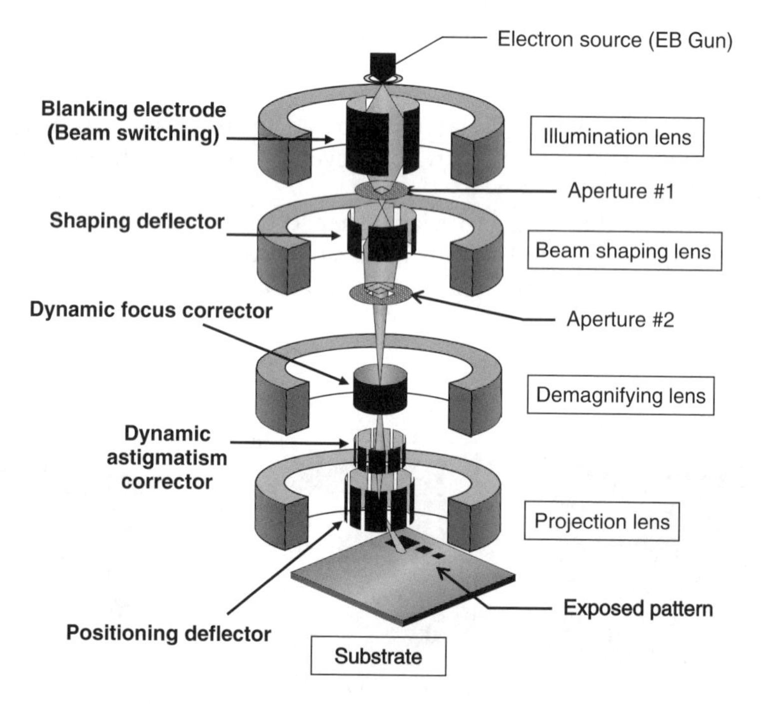

Figure 4.19 Typical electron beam system column

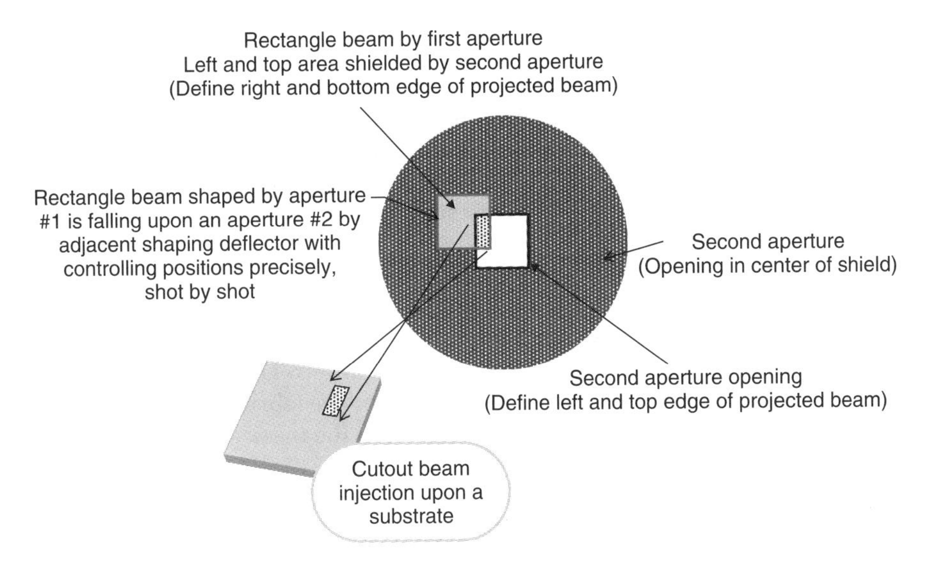

Figure 4.20 Variable shaping of electron beam

is fabricated. (x) The stage is shifted to the next field and the same process begins again to stitch adjacent patterns.

Figure 4.20 describes the process of obtaining the desired rectangle pattern. A square pattern generated at the first aperture is shifted to the second, and then exposed through the second aperture. The beam shaped in each target form is emitted at the desired location in the field, one after another. This is a vector scan method, using a variable-shaped beam (as indicated in Figure 4.18(c)). Compared with the point beam, the variable-shaped beam is inferior in terms of resolution, but a relatively fine pattern can be generated with high speed and fidelity. This is an important technology that has been put to practical use for photomask writing in the semiconductor industry (Table 4.1).

The function of the spot beam is identical, but it skips the process of shaping the beam into a rectangular form.

Electron beams are not only used to make photomasks and exposure master plates, but also exposed directly on an Si substrate at the stage of semiconductor development (called direct writing). However, since an EB can only expose a limited number of wafers in a given amount of time, it has been replaced by a photomask projection tool illuminated with light at the volume production phase.

4.2.3.3 Mold Pattern Fabrication

Generally, an EB writer with vacuum chamber is used to expose the EB to a resist on a plate/work piece. A substrate coated with resist (EB resist) is loaded onto the stage of the tool and exposed with the stage moving. Then the substrate is unloaded from the tool and developed. The pattern formed from the EB resist is used as a mask to etch membrane or material under the resist layer.

By using EB lithography, two options are available for fabricating semiconductor devices. The first option is called direct writing – direct fabrication of semiconductor integrated circuits on a wafer. The second option uses photomasks. In order to fabricate a photomask, a pattern is formed from a thin membrane including Cr, etc. on a QZ glass plate. The fabricated photomask is loaded onto a semiconductor exposure tool, and the mask pattern is exposed onto a semiconductor wafer at the production line.

Similarly, the mold fabrication process has two options. Figure 4.21 shows a flow chart for the mold pattern fabrication.

In addition to wafer direct writing, direct cutting of the substrate QZ glass is also an option for mold fabrication.

In photomask patterning, a resist with minimum but necessary thickness for etching a thin uniform Cr layer on the QZ substrate is required.

In mold fabrication, the resist should be thick and durable enough for the required etching depth. That is because, in order to obtain a mold with a 3D shape, it is necessary to fabricate substrate material made of Si or similar under the resist pattern on a substrate. For the selection of resist, one with high resistance to dry etching plasma is recommended.

Generally, the etched depth on a substrate will be uniform if the resist pattern has the same thickness after the development process. In contrast, to efficiently obtain a topographical shape by dry etching, some creative exposure methods will be required to make a 3D resist profile after development. To achieve such a 3D profile, the data structure of the pattern data needs to correspond to each location of the figures required for the target resist profile. Although 3D pattern drawing is typically more complex and takes longer for each pattern than writing of photomasks, it may be a shortcut to realize a target shape with a simpler process and take less time for some applications. Figure 4.22 shows an example. It takes a very long time to make a fine 3D pattern using EB lithography, while laser beam fabrication makes it easier although this is applicable only to a limited larger pattern size.

Applying the resist process for EB exposure requires taking into account the substrate matrix, work-piece materials to be etched, cross-section profile for resist pattern, and target mold configuration. Details on the selection process are described in Section 4.2.2 above.

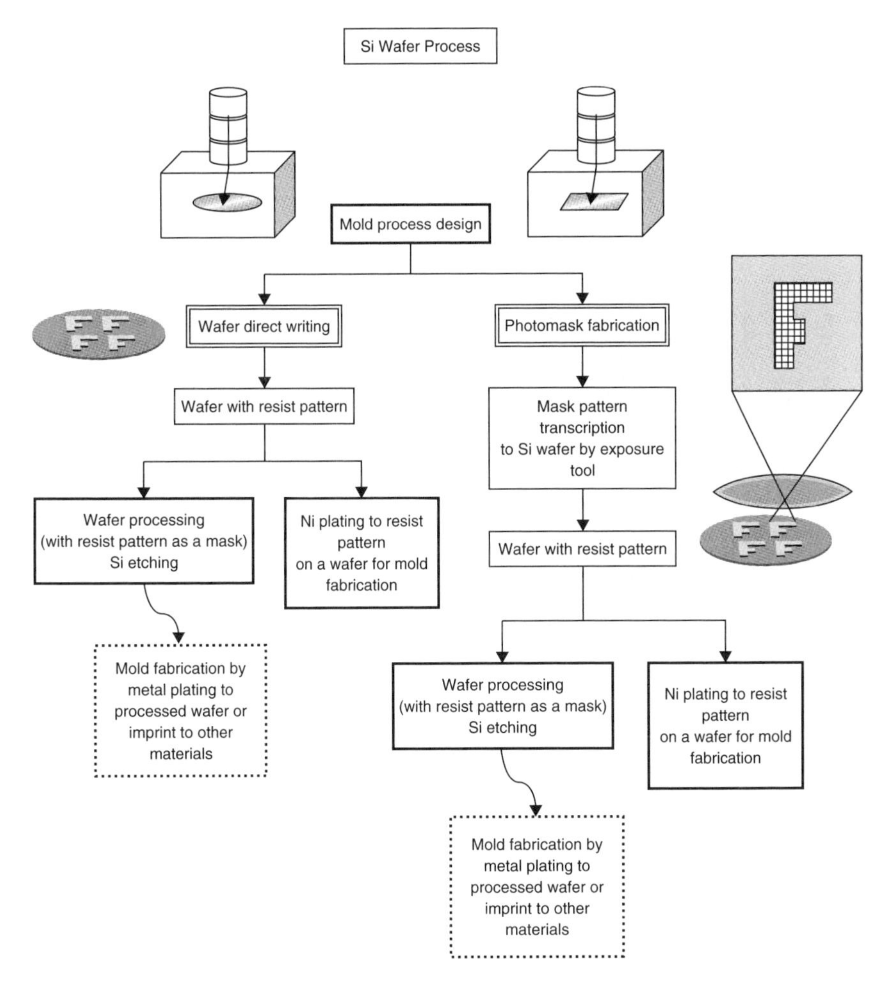

Figure 4.21 Si mold fabrication processes

4.2.3.4 Blanks Configuration – Preparation for Writing Process

In this section, essential factors for the EB exposure process will be described [7].

(a) Sensitivity of resists
An essential factor for the EB resist is its sensitivity to electron exposure against given development conditions for each resist. Some types of resist

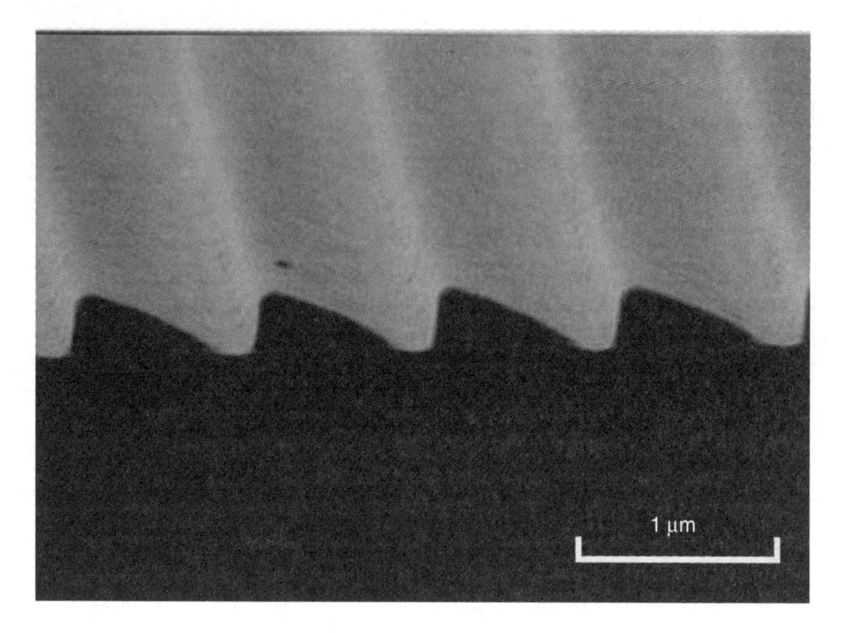

Figure 4.22 3D pattern of resist. *Source*: Courtesy of ELIONIX

react to a small number of electrons, while some types of material require a large amount of exposure for reaction and a resist pattern to be formed.

In the case of a highly sensitive resist, the writing speed is faster and the writing time is shorter. However, a material with high sensitivity to electrons also has the possibility to deteriorate patterns with unwanted electrons. A resist with a relatively low sensitivity tends to be suitable for high-resolution patterns. And the thickness of the resist affects the resist sensitivity in practice. A thinner resist tends to have a shorter development time because it has only a small number of electrons. Also, there is a rule that the sensitivity of a resist falls to about half when the acceleration voltage of the EB doubles.

Table 4.3 gives examples of typically used resists. The chemically amplified resists shown in Table 4.3 are highly sensitive, having an amplification function of electron sensitivity radicals by incident beams.

(b) Resolution

The resist pattern is formed due to the differential rates between the development speeds of the electron-exposed area and the unexposed area. A narrow-focused EB is effective to fabricate fine patterns of nanometer order. While beam focusing depends on the performance of

Table 4.3 Advanced electron beam resists

Resist	Resist Tone	Sensitivity ($\mu C/cm^2$)			Operation energy @ kV	Chemical type	Resist Developer	Manufacturer
		1 - 10	- 100	- 1000				
FEP-171	positive	✓			50	Chemical amplified	TMAH	FUJI FILM Electronic Materials
FEN-27X	negative	✓	✓		50	Chemical amplified	TMAH	FUJI FILM Electronic Materials
SEBP series	positive	✓	✓		50	Chemical amplified	TMAH	Shin-Etsu Chemical
SEBN series	negative		✓		50	Chemical amplified	TMAH	Shin-Etsu Chemical
OEBR-CAP series	positive		✓		50	Chemical amplified	TMAH	TOKYO OHKA KOGYO
NEB22	negative	✓			50	Chemical amplified	TMAH	Sumitomo Chemical
SAL-601	negative		✓		50	Chemical amplified	TMAH	Rohm and Haas
ZEP-520A	positive			✓	50-100	Conventional	Organic solvent	ZEON CORP

the EB column, a high voltage of EB acceleration generally makes for a narrow beam. Then, a pattern formation order of 10 nm will be achieved with an acceleration voltage of 100 kV, as shown in Table 4.1. At the same time, accelerated electrons with higher energy tend to penetrate the resist material and come to reduce the reaction. Thus, the sensitivity will be lowered as described above.

In some instances, electrons penetrating the resist layer will be reflected by the underlying materials. Then the reflected electrons return to the resist, and deteriorate the pattern resolution. The electron dosage needs to be tuned for the content of heavy atoms in the lower layers. It is possible to obtain high resolution patterns with a less sensitive resist which is less affected by small number of electrons or low energy. The ZEP-520A resist listed in Table 4.3 is a standard resist for nanoimprint mold fabrication. It has low sensitivity and can be used for less than 10 nm resolution.

From a microscopic point of view, there are minute peripheral irregularities in the line sidewalls of the resist pattern. This is referred to as "line edge roughness" (LER), and considered to be caused by granularity structures in a resist or resist process. Line edge roughness should be controlled, depending on the intended use.

(c) Dry etching tolerance

In succession to the EB exposure process, the resist pattern is formed by a developing process. Regularly, as an etching mask, the resist pattern will be used for dry etching utilizing gas plasma. The underlying materials will be removed by the dry-etching process. The gas generated from the dry-etching process will also trim away the masking resist pattern. Therefore, the resist material is required to have dry-etching resistance. In order to achieve a target depth by etching, the resist should have sufficient thickness to compensate for the expected reduction by etching. While the resists listed in Table 4.3 have high durability for dry etching, the resist should be selected appropriately considering the etching process conditions, purpose of fabrication of the material, etc.

(d) Fabrication of EB lithography

The first step is to determine whether to make a photomask, expose to an Si substrate, or manufacture a mold by dry etching a synthetic QZ plate using the EB exposure method.

Specification of the pattern size – whether it is fine or not – and the area size is also required. Mostly, it is essential to determine the etching method to fabricate the designed pattern. Then, in accordance with its objectives, the substrate, exposure tool, resist type, and etching equipment need to be coordinated. This requires sophisticated coordination of the EB exposure dose, resist thickness, and development methods. As the resist film becomes thinner, the resolution will be improved; however, an adequate resist thickness is required to obtain the targeted etching depth where a resist layer is used as a mask. If a sufficient thickness is not available, sometimes a layer with stronger resistance to dry etching is inserted between the resist layer and the substrate. For example, when etching a glass substrate, a Cr layer less than 0.1 μm thick is used for the mask. When the etching depth is as small as several hundred nanometers, a Cr layer of around 0.01 μm is enough.

The EB resist needs to satisfy high resolution as well as dry-etch tolerance requirements. However, this kind of resist type is often insensitive to an EB exposure that would take a longer writing time, which in turn leads to a higher cost. Therefore, the development of a resist process is most important for maximizing resolution while minimizing process cost.

In cases where a 3D resist pattern will be formed after developing the resist film layer, the resist material needs to react in proportion to the quantity of electrons received. At the same time, the resist layer needs to be shaped with its height relative to the quantity of electrons received. Although chemically amplified resists listed in Table 4.3 have high sensitivity and high resolution, they are not suitable for 3D patterning since they have high selectivity. In fact, a rather conventional resist type is suitable.

Preparation of data for the EB writer includes pattern data defining the pattern design and job-deck data containing writing instructions and/or tool conditioning. CAD formats, such as GDS-II or DXF, are commonly used as pattern data format for EB lithography.

4.2.4 Dry Etching

4.2.4.1 Principles of Dry Etching

As described above, the dry-etching technique is employed for QZ and Si mold fabrication. Discharged plasma generated from mixed gases including fluorine or chlorine atoms is generally used for etching of these materials. Many reactive species like electrons, ions, and radicals exist in such plasma. They cause chemical reaction on the substrate surface, and then volatile reaction products – for example, SiF_x, Si_xCl_y, CO_x, H_2O – are removed from the surface as gas phases.

Figure 4.23 illustrates a widely used inductively coupled plasma–reactive ion etching (ICP–RIE) system. The reaction gas is supplied into a vacuum chamber. Radio Frequency (RF) Power in the range of several hundred watts is applied to a coil wound on the outside of the upper part of the vacuum

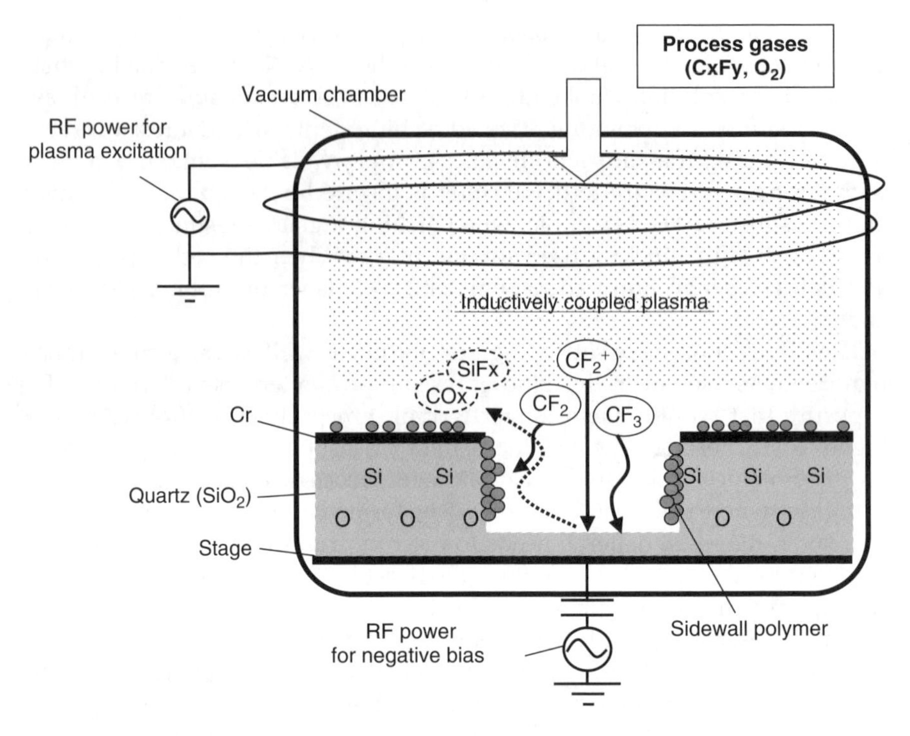

Figure 4.23 Schematic diagram of etching phenomenon in ICP–RIE system

chamber. Continuous changes in coil current generate a magnetic field in the vacuum chamber, and then an electric field in the gas is generated from the changes in that magnetic field. This electric field causes electron desorption from the gas molecules and also accelerates the movement of electrons. Accelerated electrons collide with other particles and cause continuous desorption of electrons. As a result, a high-density plasma is generated in the reaction chamber.

The generated plasma contains many excited radicals. Etching occurs as a reaction between highly reactive radicals and atoms on the substrate surface. However, such reactions are not fast enough as they stand, and such an etching could proceed toward not only the vertical but also the horizontal direction. In order to solve these problems, RF power should be applied to the substrate stage to generate negative bias voltage. Positive ions in the plasma should be vertically sputtered toward the surface of the substrate. The kinetic energy of ions contributes to the progress of etching by assisting the radical reaction on the surface and the desorption of the reaction product.

In contrast, some species in the plasma (e.g. radicals, ions, and molecules which create CF-based polymers, Cr sputtered from the surface, etc.) inhibit the etching reaction by depositing on the bottom and sidewalls of the etched aperture. However, as described above, other incident ions with high energy promote the removal of such deposits and prevent them from accumulating on the bottom. Thus, etching proceeds on the bottom of the aperture. Meanwhile, incident ions with high energy propagate in a direction parallel with the sidewall and consequently, they fail to provide energy to remove the deposit from the sidewall. The deposit then accumulates on the sidewall and forms a protective layer. This layer is commonly called the "sidewall passivation layer" or "sidewall polymer." The sidewall passivation layer curbs the horizontal etch rate and realizes anisotropic etching toward the vertical direction.

4.2.4.2 Control of Etching Profile

In a typical nanoimprint mold fabrication, dry etching is employed to obtain a vertical sidewall shape in the patterns as shown in Figure 4.24(a). However, under some etching conditions the technique fails to achieve such an ideal shape and instead causes various shape changes as shown in Figure 4.24(b)–(f).

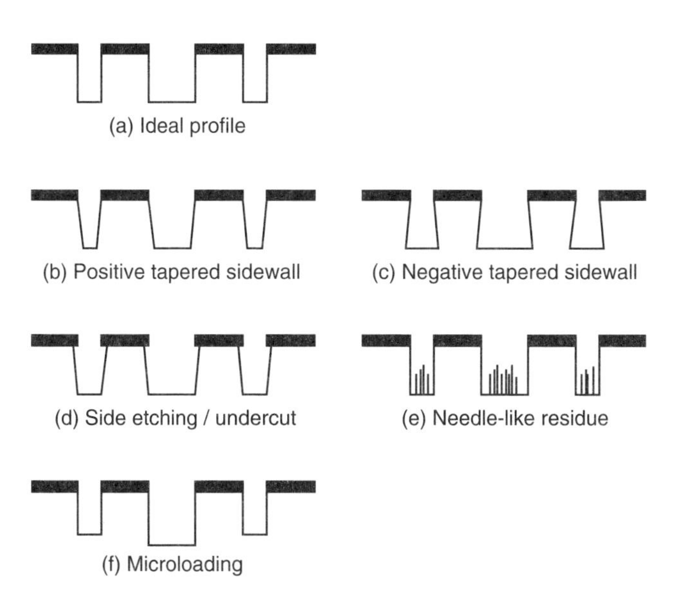

Figure 4.24 Various etching profiles

Figure 4.24(b) and (c) illustrate positive and negative tapered sidewalls, respectively. As described in Section 4.2.4.1, it is important to form a sidewall passivation layer through dry etching to obtain vertical sidewall shapes. It is necessary to balance the formation against etching of the passivation layer. Typically, any excessive accumulated sidewall passivation layer tends to cause a positive tapered sidewall, while a lack of passivation layer accumulation tends to cause a negative tapered sidewall. A slightly positive tapered sidewall could be recommended in some cases because it contributes to making mold detachment easy during imprinting. However, the bottom width becomes narrower with the progress of etching in this case, so there is a necessity to take special care in case of deep, narrow trenches and holes, because etching is stopped if both sidewalls meet at the bottom. In contrast, since negative tapered sidewalls have an obvious disadvantage when detaching the mold, they should be avoided as far as possible.

Figure 4.24(d) illustrates an example of side etching in which the etching proceeds unnecessarily in a horizontal direction, which expands the aperture width from the initial mask pattern. This is due to a lack of anisotropic etching effect in the vertical direction, with isotropic etching dominant. The etching conditions must be adjusted in this case to enhance anisotropic etching in the vertical direction. If the side etching width is small, it is possible to correct the error by narrowing the aperture of the initial mask pattern. However, if the side etching width becomes larger than half the expected trench or hole width, it is impossible to correct the error because the corrected aperture width of the initial mask pattern will become zero or less. Furthermore, given the fact that the narrower aperture size of the resist pattern makes the EB lithography process more difficult, it is desirable to minimize the side etching width as far as possible.

Figure 4.24(e) illustrates a micro needle-shaped residue formed on the bottom of the etching part. This is probably caused by micro particles with etching-resistant characteristics adhering to the bottom for some reason during the etching process, and inhibiting the adhered area of the particles from etching while etching proceeds in other areas. An extremely high density of this kind of residue could stop etching. The problem shows a tendency to become serious under conditions favorable for accumulating the protection layer (including containing a large amount of C atoms in the gas, the substrate being cooled down, etc.). Since a contaminated etching chamber also tends to cause this problem, daily management of the etching tools to maintain chamber cleanliness is very important.

Figure 4.24(f) shows a situation called "microloading," which refers to the case where the etch rate decreases in a narrow-width aperture of the pattern. It is more difficult for the radicals and ions required in etching to reach the bottom in narrower-width apertures. Even if they manage to reach the bottom and successfully trigger a chemical reaction, there is a strong probability that the reaction product will fail to release itself outside the trench and so will stay within the trench. This phenomenon leads to a local saturation of the reaction product density and an obstruction of the subsequent reaction. This mechanism possibly causes a slower etching speed in deeper trenches or holes. Generally, the narrower aperture width tends to cause a slower etching speed, but the tendency is more apparent in hole patterns than trench patterns. Hence, it is extremely difficult to achieve uniformity in depth across the whole pattern, with various sizes and shapes.

As described above, dry etching could cause various changes in etching profile. The impact cannot be ignored, especially in etching molds with very fine patterns because that could lead to significant problems. These situations actually do not occur individually but in combination, and thus could lead to more complicated irregular shapes in many cases. In order to avoid these issues as far as possible, it is necessary to tune the various etching conditions carefully, for example, the composition and ratio of gas induced in the chamber, process pressures, and ICP/bias power. In contrast, it could be necessary to purposely apply conditions that cause irregular shapes, for example, to obtain positive tapered sidewalls. Since nanoimprint molds require various structures in the etching profile other than a photomask, various etching conditions are required for mold fabrication.

4.2.5 Examples of Fabricated Mold

4.2.5.1 QZ Mold

To achieve the minimum pattern size, we focus on developing the QZ mold. In terms of pattern position accuracy and throughput, a UV nanoimprint must be developed as an alternative to lithography for semiconductor manufacture and applications to patterned media (which both require especially fine features), and such a nanoimprint technology needs a QZ mold to realize the high-accuracy performance required.

Figure 4.25 shows a QZ mold pattern with a width of about 50 nm fabricated with a variable-shaped beam EB writer and non-chemically amplified resist. By applying the photomask manufacturing process, various mold patterns with a width of less than 100 nm were achieved without significant changes to the conventional process as shown in Figure 4.25. Thus, this method could make it possible to develop molds at relatively low cost. However, the more minimized patterns have a stronger tendency to cause differences in etching

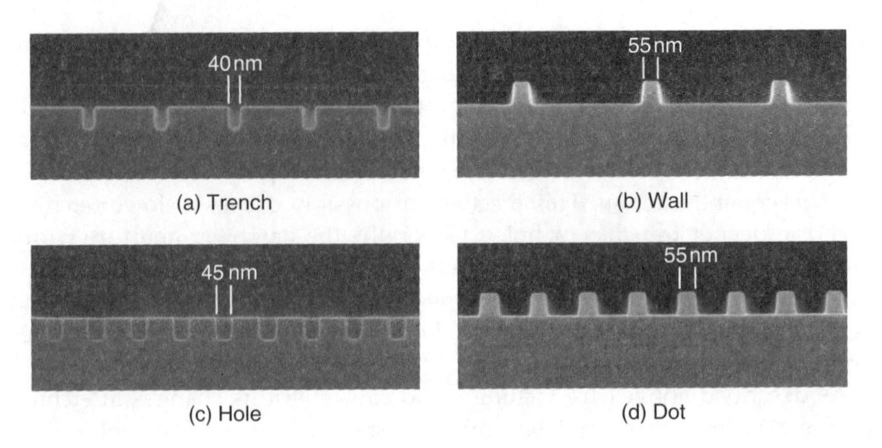

(a) Trench (b) Wall

(c) Hole (d) Dot

Figure 4.25 QZ nanostructures of width about 50 nm; both concave and convex shapes

(a) M2 width: 600 nm, Via. width: 360 nm (b) M2 width: 190 nm, Via. width: 45 nm

Figure 4.26 Trial QZ mold structures for dual damascene process

profiles among various shape patterns. In particular, this tendency is more apparent in etching depths and taper angles. So, it is not easy to fabricate a mold that has various patterns in sizes and shapes with high accuracy.

Figure 4.26 shows a simulated prototype mold for a dual damascene process. In the applied semiconductor fields it is expected that a nanoimprint will be applied [8, 9]. This mold is designed for imprinting both Via. and M2 layers in a single step. The mold basically has a structure where a Via. layer with pillar pattern is on an M2 layer with wall pattern. In order to fabricate a mold with this kind of structure, it is necessary to repeat the EB lithography

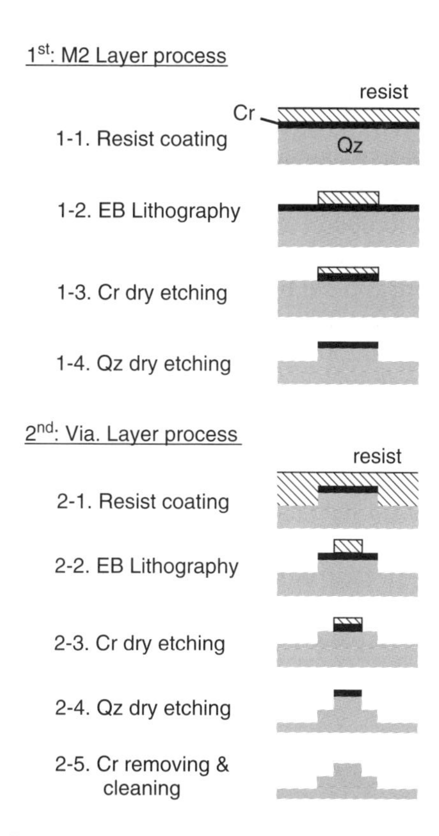

Figure 4.27 Fabrication process flow of the molds in Figure 4.20

and dry etching twice. Figure 4.27 shows an example of the manufacturing process.

Here we have introduced a process forming an M2 layer first, but another process forming a Via. layer first has also been proposed. And there is no rule on which process should be applied. No matter which process is selected, the lithography process for the secondary layer formed (the Via. layer in the case of the example in Figure 4.27) is very difficult. This difficulty is caused mainly by the necessity of forming nanopatterns on a non-flat surface. In addition, the Via. and M2 features must be aligned with extremely high registration accuracy, which is not always possible entirely at the present time. In fact, mis-registration to the left is found in Figure 4.26(b). It is very important for practical use to reduce such mis-registration.

4.2.5.2 Si Mold

Since Si microfabrication makes it possible to dry etch on a resist mask with high selectivity ratio, Si is mostly used to fabricate molds with high-aspect-ratio structures. Figure 4.28 shows mold pattern examples in which tall pillar structures are closely arranged. When fabricating such a high-aspect-ratio structure, it is critical to obtain a vertical sidewall in the etching process so the etching conditions should be set with special care. This type of mold will be effective in the fabrication of biodevices and microfluidic devices.

Also, it is possible to fabricate a mold with positive-tapered sidewalls by, in the dry etching process, optimizing the selectivity of Si and mask or controlling the sidewall polymer formation efficiency. Figure 4.29 shows some examples: (a) 300 nm pitch trench structure, (b) 1 μm pitch needle structure. These structures not only contribute to improving the mold detachment performance during imprinting, but also could be introduced in other fields including applied optical fields such as anti-reflection structures of "moth-eye type." It is not easy to obtain such structures exactly as designed, because the etching conditions need to be determined individually depending on various shapes and sizes required. However, it seems reasonable to suppose that the Si mold is used effectively for a single-step imprinting of 3D structures because of the extent of controllability of its etching profile.

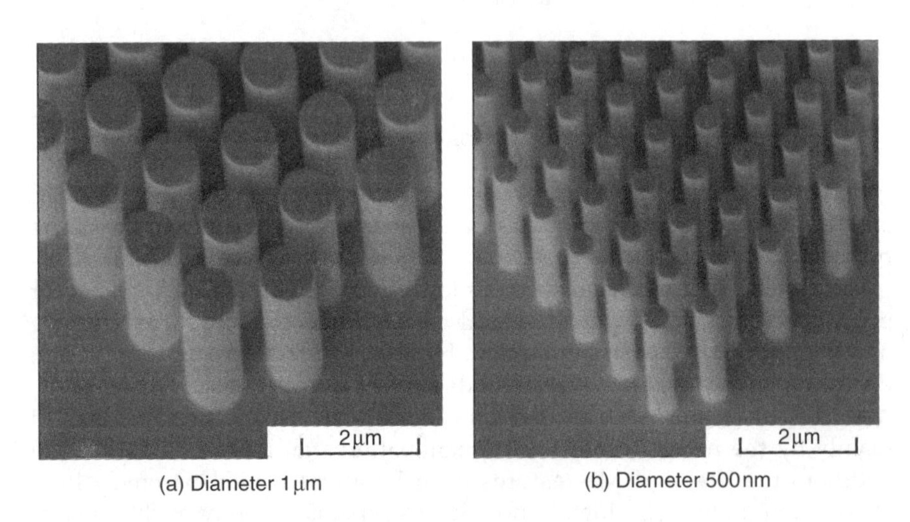

(a) Diameter 1 μm (b) Diameter 500 nm

Figure 4.28 Si mold examples of high-aspect-ratio pillars

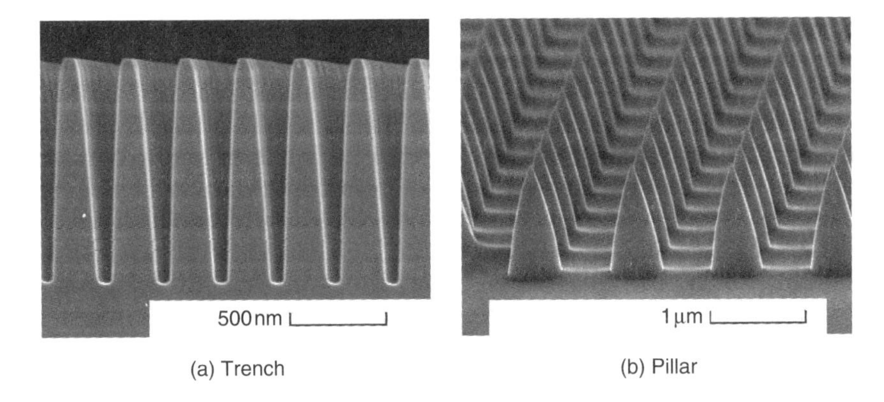

(a) Trench (b) Pillar

Figure 4.29 Si mold examples of positive-tapered structures

4.2.6 Summary

The nanoimprint technique makes it possible to fabricate nanometer-level fine patterns with very simple tools and processes. Therefore, it could reduce the cost required for patterning significantly and its application in various fields is anticipated. However, extremely high accuracy is required for nanoimprint molds in exchange for the simple imprinting process. In order to satisfy the requirement, photomask manufacturers have taken the lead in trying to apply photomask manufacturing technology to nanoimprint mold fabrication.

Photomask manufacturing has already succeeded in fabricating fine patterns with a line width of one micrometer or less, arranged within a 6 inch square without critical defects, and EB lithography and dry-etching techniques have been improved to achieve that level. Photomask manufacturers have developed these techniques further to devize nanoimprint molds with a pattern line width of 100 nm and less. In order to realize such a high resolution at low cost, they have aimed at achieving the highest resolution possible with conventional tools and materials. Meanwhile, they have also considered applying a spot beam EB writer or a new type of high-resolution EB resist, but how to achieve high throughput and cost reduction is a challenging issue.

In addition, they are not only pursuing a scale-down in pattern size but also trying to fabricate high-aspect-ratio and 3D structures. In order to resolve these issues, it has proved important to review the process conditions (which have been considered as a cause of irregular-shaped patterns) and then actively build viable process conditions to control the various etching profiles. Furthermore, some techniques apart from photomask manufacturing – including the micromachining process – have also been used to achieve various-shaped structures.

It is expected that using these developed molds will contribute to the useful development of nanoimprint technology, for example, the determination of imprinting conditions for practical use, and improved durability of molds and mold release layers. These will bring about further advancement in practical applications of nanoimprint technology. In practical applications, the improvement in user-friendliness is another important factor to consider. User-friendliness includes the standardization of mold size and shape, and the establishment of a mold cleaning method during mass production. In line with these improvements, mold manufacturers are responsible for providing higher-quality molds by reducing defects and establishing repair technology.

References

[1] Goto, K., Mori, K., Hatakoshi, G., and Takahashi, S. 1987. Spherical grating objective lenses for optical disk pick-ups. *Jpn. J. Appl. Phys.* **26**: 135.

[2] Chou, S.Y., Krauss, P.R., and Renstrom, P.J. 1995. Imprint of sub-25 nm vias and trenches in polymers. *Appl. Phys. Lett.* **67**: 3114.

[3] MacDonald, S. *et al.* 2005. *Design and fabrication of nano-imprint templates using unique pattern transforms and primitives.* Proc. SPIE, Vol. 5992.

[4] Yoshitake, S. *et al.* 2007. *The development of full field high resolution imprint templates.* Proc. SPIE, Vol. 6730.

[5] Rai-Choudhury, P. 1997. *Microlithography, Micromachining, and Microfabrication,* Vol. 1. Bellingham, WA: SPIE, chapter 2, p. 139.

[6] Rizvi, S. 2005. *Handbook of Photomask Manufacturing Technology.* Boca Raton, FL: Taylor & Francis, section 2, p. 19.

[7] Rizvi, S. 2005. *Handbook of Photomask Manufacturing Technology.* Boca Raton, FL: Taylor & Francis, section 5, p. 321.

[8] Jen, W.L. *et al.* 2007. *Multi-level step and flash imprint lithography for direct patterning of dielectrics.* Proc. SPIE, Vol. 6517.

[9] Nagai, N. *et al.* 2009. Copper multilayer interconnection using ultravaiolet nanoimprint lithography with a double-deck mold and electroplating. *Jpn. J. Appl. Phys.* **48**: 115001.

5

Ultraviolet Nanoimprint Lithography

Jun Taniguchi[a], Noriyuki Unno[a], Hidetoshi Shinohara[b], Jun Mizuno[c], Hiroshi Goto[d], Nobuji Sakai[e], Kentaro Tsunozaki[f], Hiroto Miyake[g], Norio Yoshino[h], and Kenichi Kotaki[i]

[a]*Department of Applied Electronics, Tokyo University of Science, Japan*
[b]*Department of Electronic and Photonic Systems, Waseda University, Japan*
[c]*Nanotechnology Research Laboratory, Waseda University, Japan*
[d]*Toshiba Machine Co., Ltd, Japan*
[e]*Samsung R&D Institute, Japan*
[f]*Asahi Glass Co., Ltd, Japan*
[g]*Daicel Corporation, Tokyo Head Office, Planning R & D Management, Japan*
[h]*Department of Industrial Chemistry, Tokyo University of Science, Japan*
[i]*SmicS Co., Ltd, Japan*

5.1 Orientation and Background of UV-NIL

For the next generation of nanoscale pattern transfer, a strong need exists for a nanoscale patterning technique which has a high throughput and cost-effective processing. Nanoimprint lithography (NIL) is considered to be

Nanoimprint Technology: Nanotransfer for Thermoplastic and Photocurable Polymers, First Edition.
Edited by Jun Taniguchi, Hiroshi Ito, Jun Mizuno, and Takushi Saito.

a major breakthrough for this next-generation lithography (NGL) because of its high resolution and simpler process compared to conventional lithography technology. The basic idea of NIL was investigated by S. Fujimori (1976, NTT, Japan) and it was termed the "mold mask method" [1–3] at that time (Figure 5.1). The mold mask method encompasses approximately the same method as that of the contemporary NIL process. For example, replicated molds were transferred from an original mold, which was fabricated by electron beam lithography or focused ion beam. Poly(dimethylsiloxane) and negative-tone photoresist were used as transferred thermal and UV-curable resin, respectively.

Using the mold mask method, the 2 μm square pattern and 3D micro lens pattern were obtained in the 1970s. At that time, it was too early to recognize the significance of the resin viscosity, the mold hardness, and the cushioning mechanism for uniform transfer process. However, photolithography earned its place as a major technology at that time because of its high throughput. It should be added that the confirmed resolution of this method was only of sub-micrometer order because fabrication of a fine patterned mold (less than 100 nm) was difficult in the 1970s.

However, the success of the 25 nm vias pattern transfer using NIL, reported by S.Y. Chou in 1995 [4], changed the situation regarding mold methods. Since deep UV optical lithography tool costs (for stepper or scanner) keep rising as the pattern size shrinks, the purpose of introducing NIL is to replace optical lithography. This original NIL process is shown in Figure 5.2. This process uses thermoplastic resin, so the process is called "thermal cycle NIL." Thermoplastic resin is solid at room temperature but above the glass transition temperature (T_g), this resin changes to a liquid phase. In this state, thermoplastic resin is made to flow and fill the mold pattern through the application of pressure. The thermal cycle NIL process operates as follows. First, the nanopatterned mold and resist-coated silicon substrate are prepared (Figure 5.2(1)). The resist thickness is around a few hundred nanometers or less for lithography use. Next, the resist layer temperature is elevated to over 200 °C. The glass transition temperature of PMMA (polymethylmethacrylate) is 105 °C, thus 200 °C is quite sufficient for molding (Figure 5.2(1)). Then, the mold contacts the resist layer and pressure is applied at 13 MPa (Figure 5.2(2)). At this time, the resist flows and fills up the mold structure. After the resist has flowed, a solidification process is needed. This is done by cooling the resist layer while keeping the applied pressure, thus solidifying the resist layer (Figure 5.2(2)). After the resist layer solidifies, the mold is

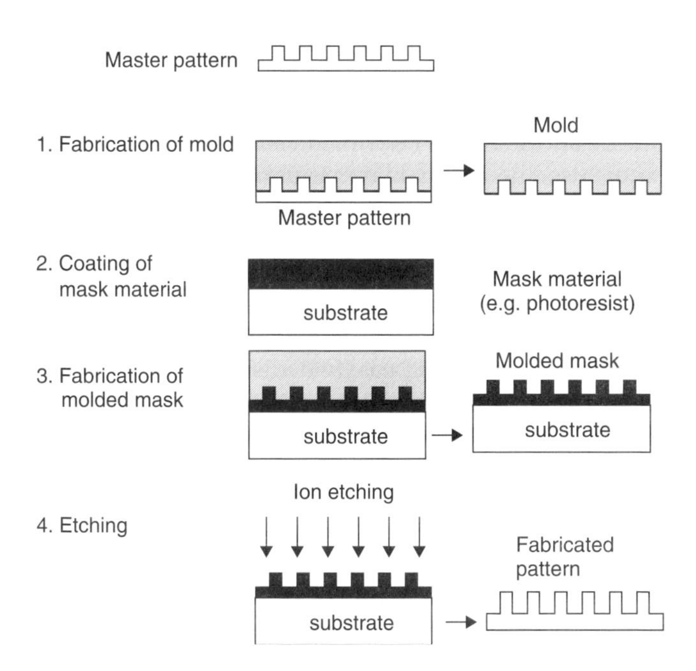

Figure 5.1 The process schematics of the mold mask method (1976)

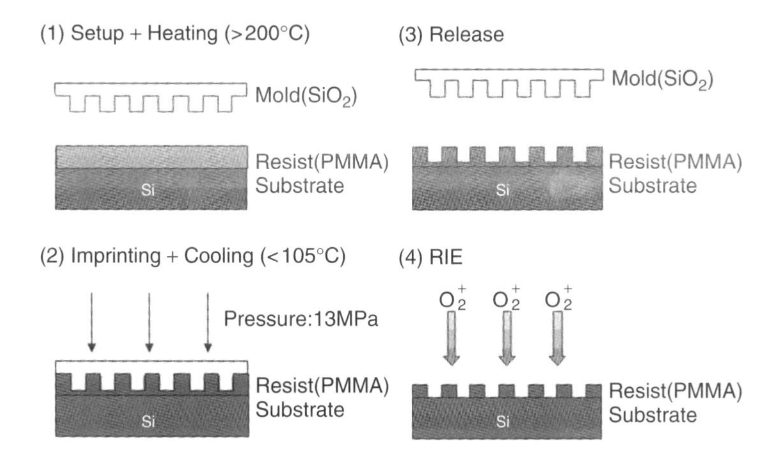

Figure 5.2 Original NIL process (4)

released from the resist layer and we can obtain the nano-order structure (Figure 5.2(3)). An imprinted resist pattern is the reverse of the mold pattern. For example, a concave pattern on the resist corresponds to a convex pattern on the mold. In addition, the imprinted resist layer has a residual layer. For lithographic use, the residual layer is unnecessary. Thus, to remove the residual layer, short time oxygen reactive ion etching (RIE) is the appropriate process (Figure 5.2(4)). After this process, the silicon surface appears and the imprinted resist pattern on the silicon substrate is in the same state as optical lithography (Figure 5.2(4)). Thus, using the NIL process, ion implantation, dry etching, and film deposition are carried out.

However, thermal cycle NIL has two disadvantages. One is that the heating-up and cooling-down processes are time-consuming. Lithography requires high throughput (more than 100 wafers/hour), thus, an improvement in throughput by shortened processing time is necessary. Another disadvantage is a change in pattern size because of different coefficients of thermal expansion between the mold material and substrate material. This problem is partially solved when using the same material for mold and substrate. However, heat application is not particularly suitable for lithography purposes. UV-NIL can solve these problems. The UV-NIL process is shown in Figure 5.3. This process uses photocurable resin which is liquid at room temperature (Figure 5.3(1)). Thus, resin flow can occur simply by applying pressure without heating (Figure 5.3(2)). The solidification of resin is carried out by exposure to UV light (Figure 5.3(2)). At that time, the mold needs to be transparent for the UV light, thus, quartz or sapphire is used for the mold material. After solidification, the mold is released from the resin (Figure 5.3(3)) and the subsequent steps are the same as for thermal cycle NIL. Usually, UV photocurable resin has highly sensitive materials, thus solidification is very fast. Therefore, UV-NIL can be used for lithography purposes because this allows a high-throughput process. In addition, UV-NIL is the candidate process for the below-22 nm half pitch era.

One principle of NIL is the replication of the plastic layer by a mold. This is a very simple principle, yet NIL is a powerful tool for nanofabrication not only with lithography but also with MEMS and so on, because NIL is a 3D pattern transfer. Optical lithography uses photochemical reactions of the resist layer. In contrast, NIL uses physical deformation of the resist layer. Thus, NIL is free from complex optical systems, lenses, and mask sets. This means that NIL equipment is relatively inexpensive. However, some problems still exist. NIL is contact lithography, so breaking of a mold and sticking of the resist layer on a mold sometimes occur. To prevent these phenomena, low-pressure processing and release coating on the mold are candidate processes. In addition, NIL is a same-size (1×) pattern transfer

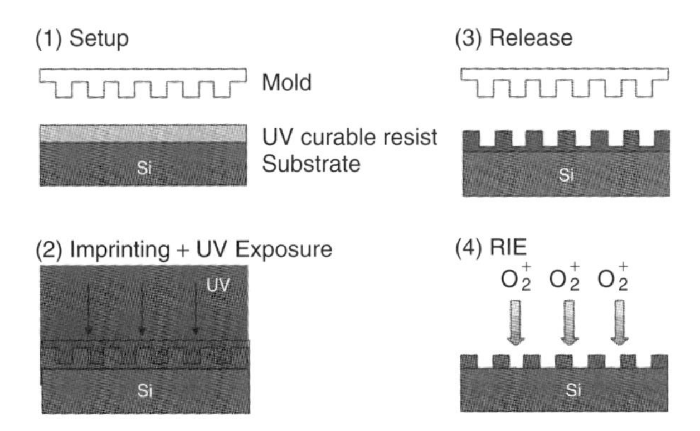

Figure 5.3 UV-NIL process

system (i.e., the replicated pattern feature size equals the mold feature size), thus, usually a sub-100 nm feature size mold is required. Fabrication of this order of mold is a very demanding process. Usually, electron beam lithography (EBL) is used for fabrication of the mold, and the EBL process requires a great deal of know-how (see Section 4.2).

In this chapter, the phenomena of UV-NIL transfer process, UV-NIL machine, UV-NIL materials, and evaluation of mold and transfer resin surface are described.

5.2 Transfer Mechanism of UV-NIL

The next generation of fine patterning requires a high-throughput cost-effective process for mass producing various devices, with a nanometer-scale pattern over a large area. UV-NIL (Figure 5.4) is a major breakthrough in this field because it offers high throughput, resolution, position accuracy, and low equipment cost.

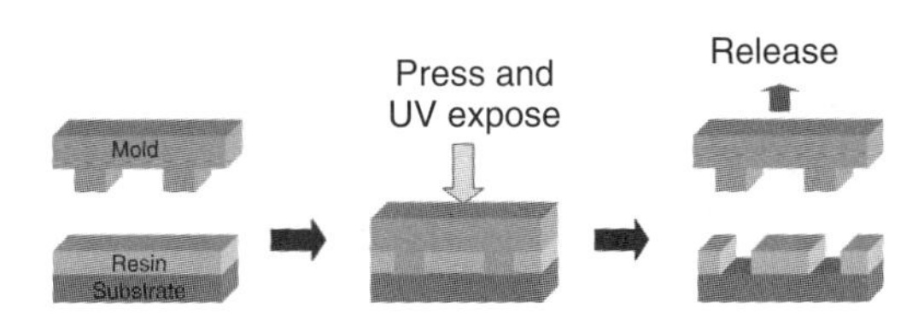

Figure 5.4 UV-NIL process

In UV-NIL, a liquid resin having a low viscosity is used, thus, the required transfer pressure is lower than that of thermal NIL. However, there are some technical challenges to be solved for practical application. In particular, it is important to elucidate the filling behavior of UV photocurable resin into the mold in order to produce an exact replicated pattern. In order to observe the filling behavior, we used a mid-air structure mold, for which air bubble defects can be ignored. As a result, the filling behavior of UV photocurable resin into the mid-air structure mold was observed clearly with a scanning electron microscope (SEM) and it was confirmed that the release agent encumbered the filling of UV photocurable resin in the nanoscale pattern. This phenomenon appeared prominently in the finer pattern.

5.2.1 Viscosity and Capillary Force

The mid-air structure molds were fabricated by control of acceleration voltage electron beam lithography (CAV-EBL) [5, 6]. Hydrogen silsesquioxane (HSQ; FOX-14 made by Dow Corning Co.) was employed as a negative-type electron beam resist. ERA-8800FE (Elionix Co.), which is a customized SEM for the delineate task, was used for the CAV-EBL system with a beam current of 6 pA. The CAV-EBL process is shown in Figure 5.5.

First, HSQ was spin-coated at 3000 rpm on a cleaned Si substrate and baked at 180 °C for 5 min, resulting in a 300 nm film. Then, EBL was carried out at a high accelerating voltage of 30 kV to delineate line patterns. Subsequently, a second EBL was carried out at a low acceleration voltage of 3 kV to delineate lines perpendicular to the first direction. Finally, the HSQ film was developed in tetramethylammonium hydroxide (TMAH 5%) for 180 s. The acceleration voltage changes the projection range of the electron beam; in this case, the EBL was carried out at 30 kV to form trough beams and 3 kV to form bridges, resulting in the mid-air structure (as shown in Figure 5.6).

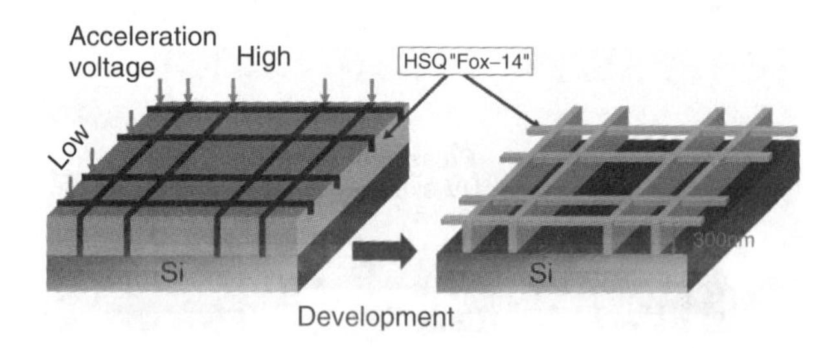

Figure 5.5 The fabrication process of the mid-air structure mold using CAV-EBL

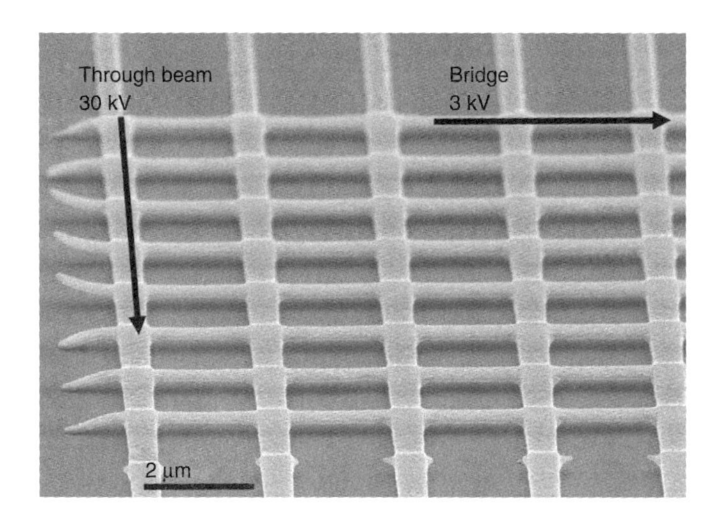

Figure 5.6 The fabricated mid-air structure mold on an Si wafer

Using the fabricated mid-air structure molds, UV-NIL was carried out at atmospheric pressure. The advantage of using the mid-air structure mold is that free channels exist under the bridges, allowing any air to escape during the transfer process; therefore, bubble defects do not occur and two adjacent bridges were considered as the both-ends-open HSQ channel (Figure 5.7).

The fabricated mid-air structure mold was coated with release agent (0.1%, Optool DSX, Daikin Industries Ltd). We used three types of UV-curable resin: PAK-01, PAK-02-TU01, PAK-02 (Toyo Gosei Co., Ltd), whose kinetic viscosities are 63.5 mPa s, 16.0 mPa s, and 9.30 mPa s, respectively, and surface tensions 30.6 mN/m, 27.1 mN/m, and 29.9 mN/m, respectively. One drop of PAK was put on the mold around the HSQ mid-air structure. A glass slide was used as a transfer substrate and the transfer pressures were 0.2 MPa, 0.5 MPa, and 0.8 MPa. The pressure hold time was 60 s, which was sufficient to complete the formation of the replicated pattern. After the hold time, UV radiation with an energy density of 4 J/cm^2 was focused onto the mid-air structure mold. Then, the mold was retracted, leaving behind a replicated pattern.

At this time, the mid-air HSQ structure was also released from the substrate in some cases. However, we assumed that the contact angle, which developed between the liquid PAK and solid mold sidewalls, was fixed during UV exposure and expressed the filling behavior. So, we observed the PAK-01 pattern and the mid-air structure mold on the glass slide after the UV-NIL process with SEM.

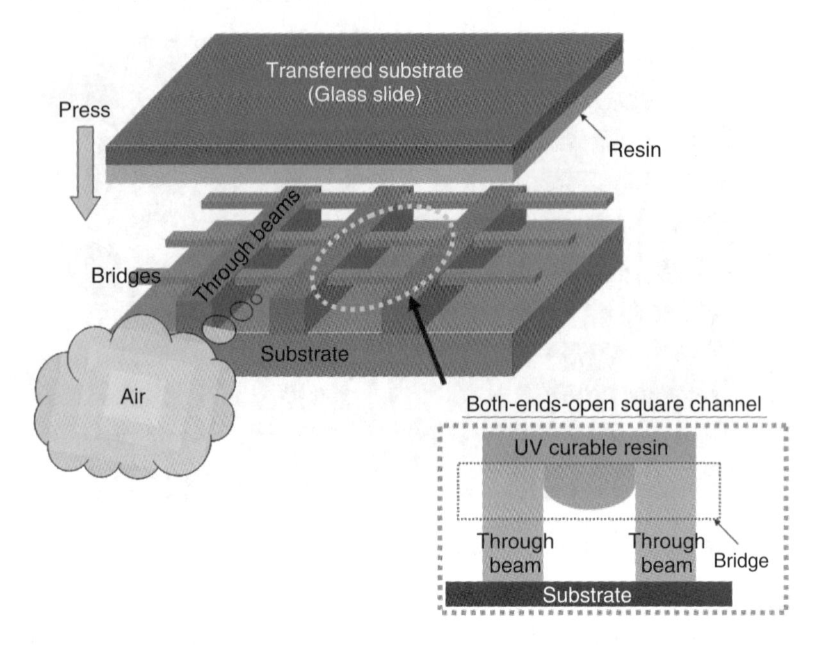

Figure 5.7 Schematic view of UV-NIL condition using the mid-air structure mold

Figure 5.8 shows SEM images of the transfer results at 0.5 MPa with the release agent. The patterned heights were 207 nm, 234 nm, and 266 nm using PAK-01, PAK-02-TU01, and PAK-02, respectively. The patterned height depended on the kinetic viscosity of the resin: the larger kinetic viscosity of the UV-curable resin tended to result in a shorter patterned height, thus, it required a higher transfer pressure.

Figure 5.9 shows SEM images of the transfer results, for varying aperture size of the mid-air structure mold. The transfer pressure was a constant 0.5 MPa and the mold was coated with the release agent. In this case, only PAK-01 was used as a UV-curable resin. When the aperture size was 500 nm, 1000 nm, and 1500 nm, the patterned heights were 72 nm, 207 nm, and 256 nm, respectively. A larger aperture tended to form a higher transfer pattern. In addition, the transferred pattern shape changed from a hemispheroid to an angular shape, as shown in Figure 5.10.

Figure 5.11 shows the relationship between the mold aperture size and the transferred pattern height by varying the transfer pressure. The pattern height with aperture size of 1000 nm and pressure of 0.2 MPa was lower than that with 0.5 or 0.8 MPa. In contrast, each pattern height with 1500 nm

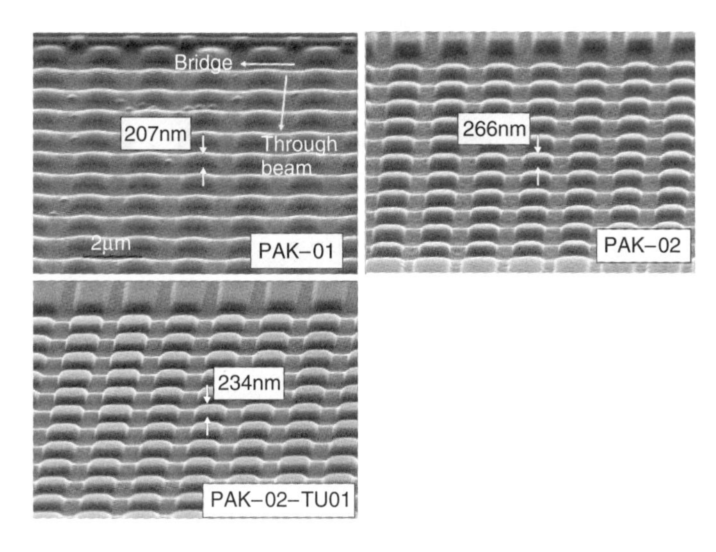

Figure 5.8 Effect of kinetic viscosity of resin on the transfer results

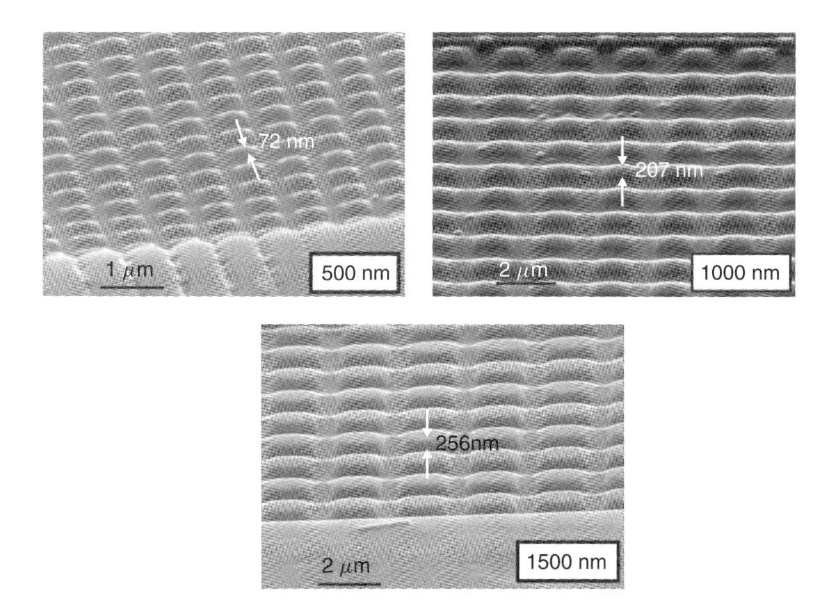

Figure 5.9 Effect of aperture size of the mid-air structure mold on the transfer results

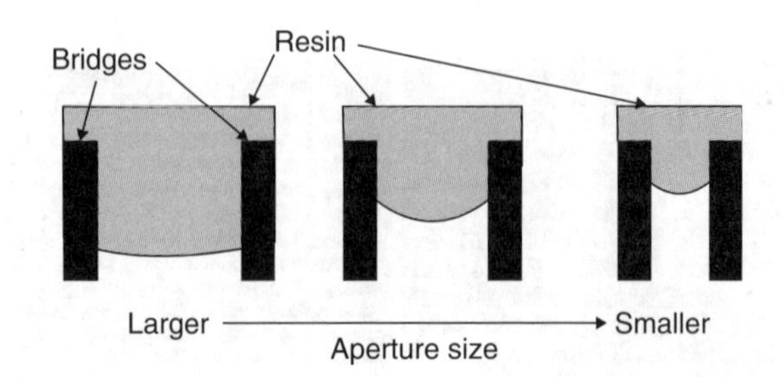

Figure 5.10 Schematic cross-sectional view of the channel in the mid-air structure mold

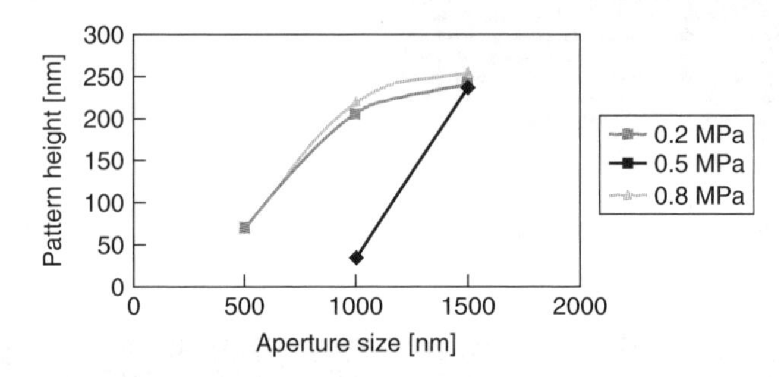

Figure 5.11 Relationship between aperture size and transferred pattern height

aperture size was approximately equal to the initial HSQ thickness. This means that the transfer pressure, which is required to fill the mold pattern with resin, depends on the aperture size, and a smaller aperture size needs a higher transfer pressure.

5.2.2 Release Coating and Evaluation of Release Properties

In the NIL process, the durability of a mold is very important to improve productivity at low cost. Although a very fine pattern can be obtained using NIL, the mold is at risk of destruction and clogging with resin because NIL is a contact process. In order to prevent these problems, a NIL mold is usually treated with a release agent. Typically, there are two main types of release coating. One is a hard material coating, such as diamond-like carbon (DLC)

[7]. DLC has high hardness, low surface energy, and low friction, thus, it is suitable as a release agent. Recently, fluorinated DLC (F-DLC) has attracted considerable interest because of its better release property due to the low surface energy of fluorine [8]. However, it is difficult to form a thin DLC layer. The other type is a self-assembled monolayer (SAM), such as a silane coupling agent, which has a hydrophobic foot with fluoride [9]. SAM is able to coat a mold with a very thin monolayer. Hence, SAM is commonly used as a release agent for molds with sub-100 nm pattern. This section mainly describes the SAM release agent and its coating method.

A silane coupling agent is one of the most popular materials of SAM and it has an organic functional group, such as alkoxysilane or chlorosilane. The reactivity of chlorosilane compound is higher than that of alkoxysilane. However, chlorosilane compound generates hydrochloric acid as a result of hydrolysis, and so must be handled carefully. Figure 5.12 shows the reaction of the alkoxysilane, which is typically used for silane coupling agents for NIL. First, the alkoxysilane group is hydrolyzed and connects to hydroxy groups on the substrate by hydrogen bonding. Then, the substrate is heated and the alkoxysilane groups form covalent cross-linkages. As a result, the surface of the substrate is covered with a monolayer of hydrophobic foot, which is composed of fluoride. In order to connect SAM to the substrate uniformly, it is very important to form hydroxyl groups on the substrate. For instance, the combination of diluted hydrofluoric acid and piranha solution (H_2SO_4:$H_2O_2 = 1:4$) treatments forms hydroxyl groups on a silicon substrate [10].

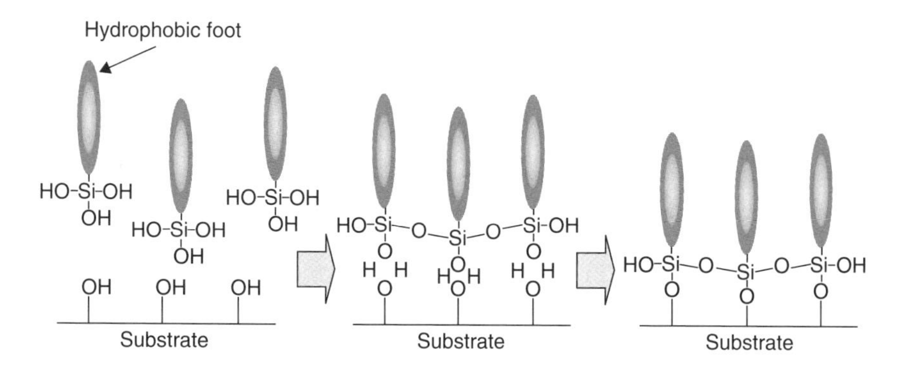

Figure 5.12 The reaction of the silane coupling agent

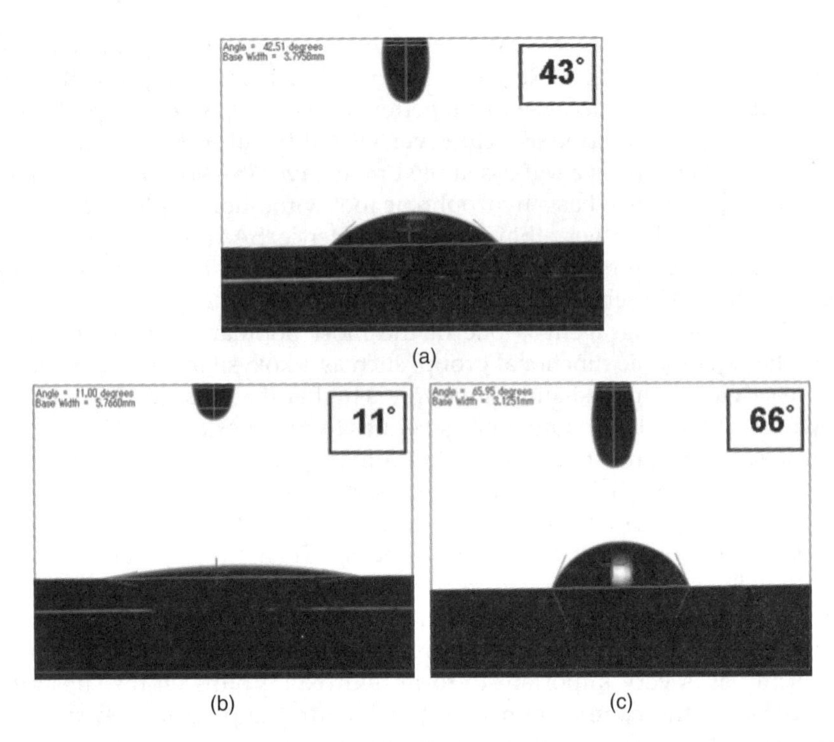

Figure 5.13 Contact angle between PAK-01 and the HSQ film: (a) initial HSQ surface, (b) after post-bake, (c) after coating the release agent

In order to evaluate the efficacy of a release agent, a contact angle is generally used as an index because the mold surface energy can be calculated from the contact angle [11]. The lower the surface energy of the mold, the smaller the peel force from the resin.

Figure 5.13 shows the contact angle between PAK-01 and the HSQ film on a silicon wafer with or without the release agent, which was measured using a contact angle measurement system (FTA125, First Ten Angstroms, Inc.). Optool DSX (Daikin Co.), which is a fluorine–silane coupling agent and diluted at 0.1% with perfluorohexane, was used as a release agent. The coating method was as follows: first, HSQ was spin-coated at 3000 rpm on a cleaned Si substrate and baked at 180 °C for 5 min, resulting in a 300 nm film. Next, the substrate was dipped into the release agent for 1 min at room temperature. Then, the substrate was heated to 100 °C in air to evaporate the solvent and help covalent cross-linkage of the alkoxysilane groups. Three conditions of the HSQ surface were investigated, with or without the release agent, and only post-baked at 350 °C.

Without release agent, the contact angle was 43° and it was increased to 66° by the release agent Optool DSX. In contrast, the contact angle of the HSQ film was decreased to 11° after post-baking at 350 °C because the organic constituent in the HSQ film was desorbed. The release agent leads to a higher contact angle and a low surface energy, and thus helps the release of the mold from the transferred substrate.

5.2.3 Release Coating Effect

As noted before, the mold is required to be coated by the release agent to prevent destruction and clogging with resin. The mold surface with release agent has low surface energy and it helps reduce the peel force. However, the low surface energy of the mold surface strongly repels the resin. As the required pattern width for the mold has been decreased, the force repelling the resin is considerable. In this section, the filling behavior with or without release agent is investigated and it is found that the filling behavior is closely related to the capillary force of the mold.

Figure 5.14 shows the PAK-01 shape, which is patterned on the glass slide by the mid-air structure mold with or without release agent. The transfer pressure and the aperture size were 0.2 MPa and 1000 nm, respectively.

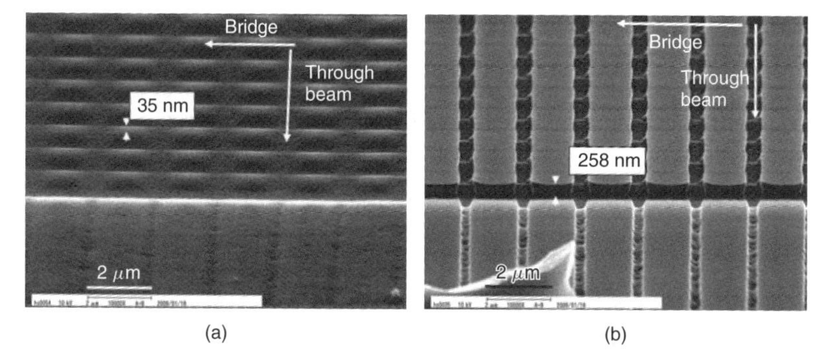

Figure 5.14 The transfer results with PAK-01 on the glass slide using the mid-air structure mold: (a) with, or (b) without release agent

Figure 5.14(b) shows that PAK-01 overflowed in adjoining apertures, since it completely filled the HSQ channel. In contrast, with release agent, PAK-01 did not overflow and the patterned height was shorter than that without release agent, as shown in Figure 5.14(a). This means that the release agent encumbered the filling of PAK-01 into the square aperture of the HSQ channel. The force, which encumbered the filling of PAK-01, is constant (P_C) and pressure surpassing P_C is required to fill PAK-01 into the aperture of the mid-air structure mold. P_C is represented by [12, 13]:

$$P_C = \frac{4\gamma \cos\theta}{a} \tag{5.1}$$

where a is the length of the channel sides, γ is the liquid surface energy, and θ is the contact angle in the channel.

In this case, the mid-air structure mold with the release agent has a negative P_C value because the contact angle θ, which is generated by the hydrophobic mold surface, is greater than 90°. Without the release agent, in contrast, the contact angle θ is less than 90°, hence, the value of P_C is positive, as shown in Figure 5.15.

For experimental determination of P_C with or without release agent, the filling behavior with respect to each transfer pressure was examined. Figure 5.16 shows the characteristics of the filling behavior of PAK-01 into the mid-air structure mold with an aperture size of 1000 nm.

The patterned height of PAK-01 without release agent was shorter than the initial HSQ thickness because of the post-baking. With release agent, the patterned height was far greater at 0.5 MPa than at 0.2 MPa. However, the patterned height without release agent remained almost constant throughout the range of pressure. This means that P_C of the mid-air structure mold, which had an aperture size of 1000 nm with release agent, was approximately 0.5 MPa. Consequently, it is confirmed that the increase of P_C due to the release agent encumbered the filling of UV photocurable resin.

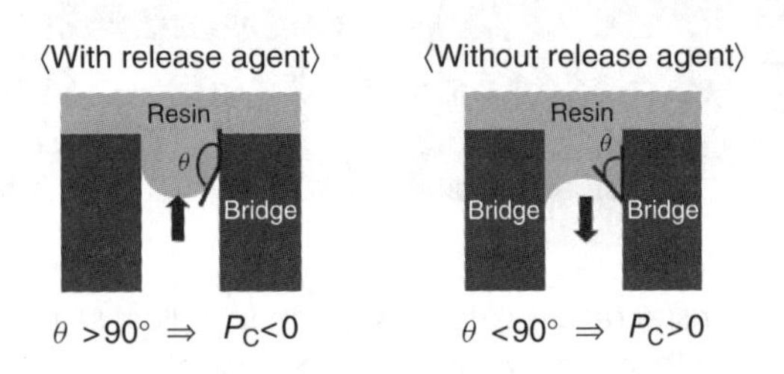

Figure 5.15 P_C dependence on the contact angle in the channel of the mid-air structure mold

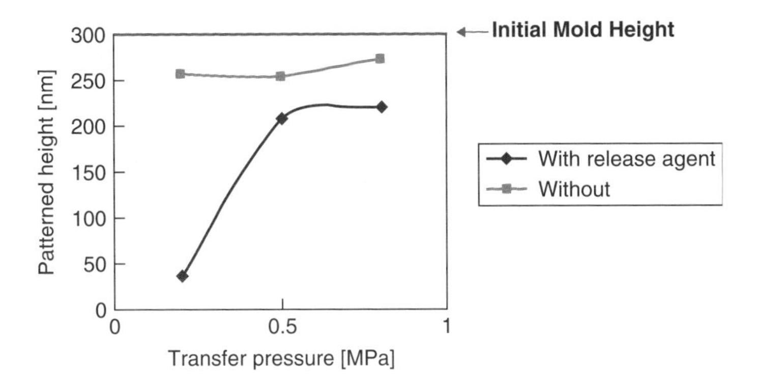

Figure 5.16 Effect of transfer pressure on filling behavior

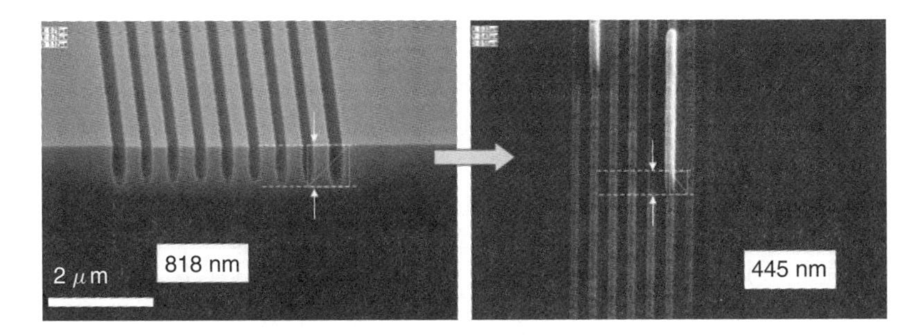

Figure 5.17 The failed transfer result at 0.6 MPa with release agent

Finally, line and space pattern molds with release agent were prepared to confirm this phenomenon with a normal NIL mold. Figure 5.17 shows SEM images of the mold with release agent and the transfer result using PAK-01 at 0.6 MPa. The transferred pattern height is significantly shorter than the mold depth. In contrast, Figure 5.18 shows SEM images of the mold with release agent and the transfer result at 1.2 MPa. Although the transferred pattern was shrunk slightly because of the curing shrinkage of the resin, the transferred pattern was approximately equal to the mold depth. Therefore, the transferred pattern using a normal NIL mold is similarly influenced by P_C.

Consequently, it is important to elucidate the surface state in the NIL mold pattern with release agent in order to be successful in the fine NIL process. Our observation method using the mid-air structure demonstrated that P_C was increased by the release agent which encumbered the filling of UV

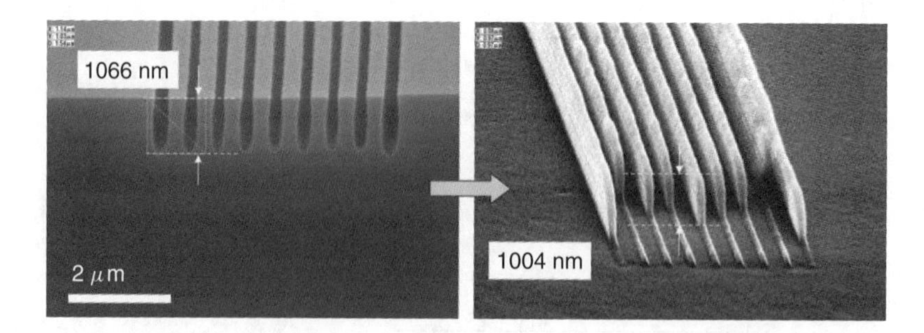

Figure 5.18 The transfer result at 1.2 MPa with release agent

photocurable resin into the mold. Further elucidation of P_C will enable mold patterns to be replicated exactly with UV-NIL.

5.3 UV-NIL Materials and Equipment

5.3.1 Ubiquitous NIL Machines

UV-NIL has attracted considerable attention as a micro-/nanopattern fabrication technology. However, the UV-NIL equipment developed thus far is large-sized and expensive. Inexpensive and compact UV-NIL equipment is desired, especially in the field of research and development. Two compact equipments have been developed: one for UV-NIL [14, 15] and the other for UV roller imprinting [16, 17].

Figure 5.19 shows the UV-NIL equipment developed. It consists of a bottom plate to fix the substrate, a top plate to fix the mold, and a weight to apply the loading force. The maximum diameter of the substrate is approximately 30 mm. UV light is incident from the top side of the equipment through a quartz window and the mold. Because conventional UV-NIL equipment is operated under atmospheric pressure, pattern defects are introduced due to the formation of air bubbles [18, 19]. To avoid the formation of these defects, a vacuum chamber is employed in this equipment.

The cavity formed between the top plate and the bottom plate is hermetically sealed and can be evacuated through a hole in the bottom plate. Figure 5.20 shows results of UV imprinting under atmospheric pressure and under reduced pressure. By imprinting under reduced pressure, the mixing of air bubbles with the resin can be prevented.

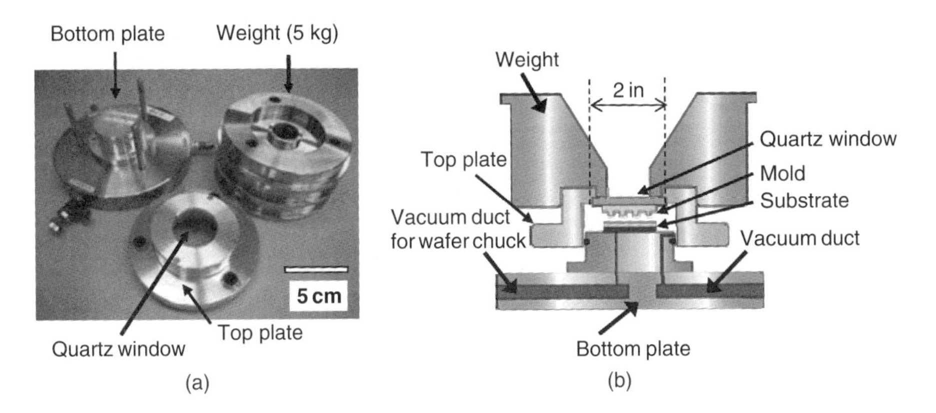

Figure 5.19 UV-NIL equipment: (a) photograph, (b) schematic

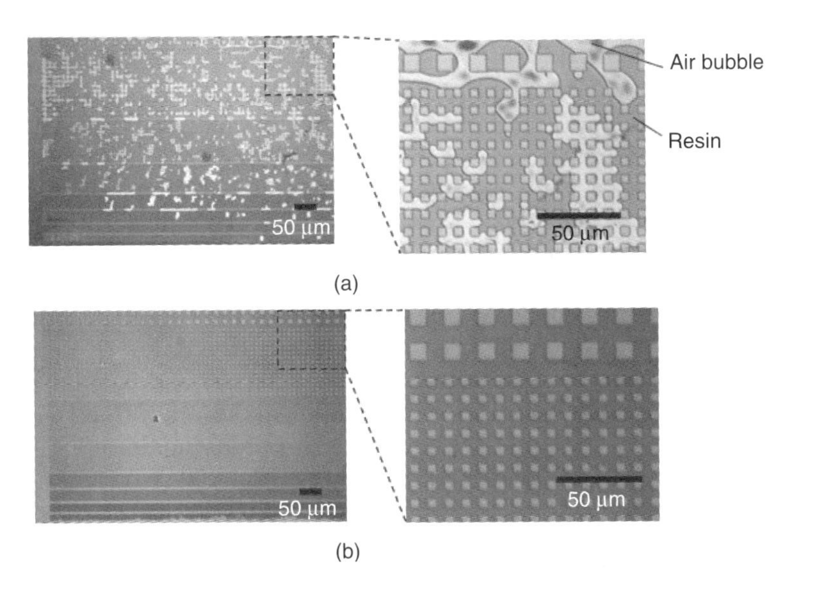

Figure 5.20 Photomicrographs of UV-imprinted samples: (a) under atmospheric pressure, (b) under reduced pressure

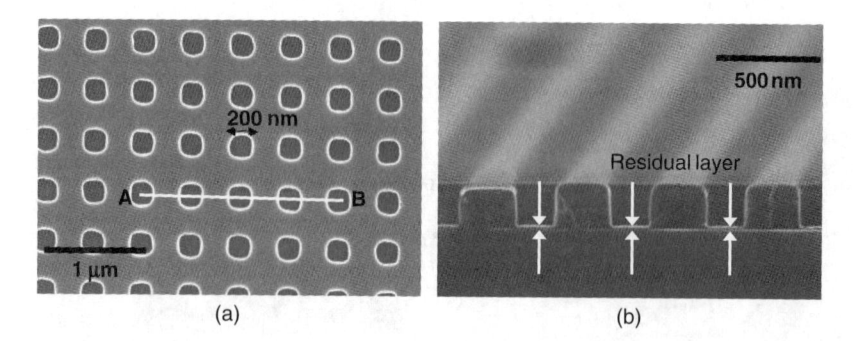

Figure 5.21 SEM image of UV-imprinted samples under optimized conditions: (a) top view, (b) cross-sectional view (along line A–B in (a))

The UV imprinting conditions were optimized by using a quartz mold. The mold had many rectangular dot patterns of diameter 200 nm, height 225 nm, and pitch 500 nm. Figure 5.21(a) shows a SEM image of the resin pattern fabricated using the above-mentioned UV-NIL equipment. A pressure of 0.25 MPa was applied under a reduced pressure of 450 hPa. After imprinting, the resin had hole patterns, diameter of approximately 200 nm, depth of 223 nm, and pitch of 500 nm. This image shows that the quartz mold patterns were imprinted successfully with a high-dimensional accuracy. Figure 5.21(b) shows a cross-sectional SEM image of the resin pattern. In the case of a resin thickness of approximately 210 nm, a residual thickness of 10 nm was measured.

Figure 5.22 shows a photograph and the operating mechanism of the equipment developed for UV roller imprinting. The roller is 32 mm in diameter and 50 mm wide. The roller can be detached and handled easily; thus, flexible film molds can be used by wrapping them around the roller. The stage can be moved 100 mm in the horizontal direction, and the replication of an area of 30 mm × 100 mm is possible. A DC servomotor with a maximum rotation speed of 3000 rpm is used. Its rotating motion is converted into a scanning motion with a speed of 200 mm/min. As the DC servomotor rotates, the roller synchronizes its motion with that of the stage and the sheet substrate is transported. Sheet substrates are commercially available as 30 mm wide and 320 mm long. Photocurable resin was first coated on the sheet substrate using a bar coater before setting the sheet in the equipment. In order to keep the tension of the transported sheet constant, a spring is included in the equipment. The UV light intensity can be controlled by varying the slit width between 1 mm and 3 mm.

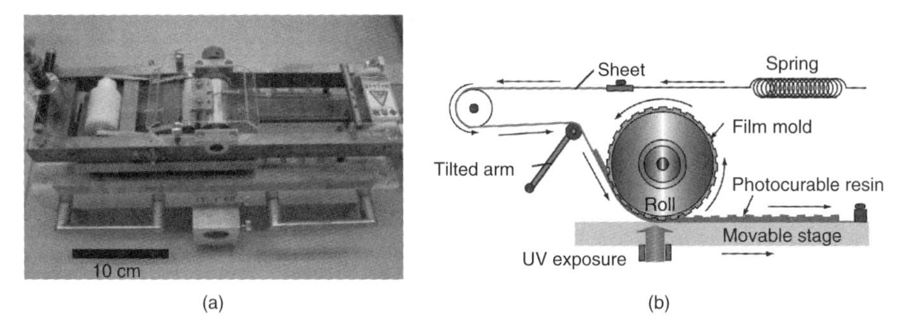

Figure 5.22 UV roller imprinting equipment: (a) photograph, (b) operating mechanism

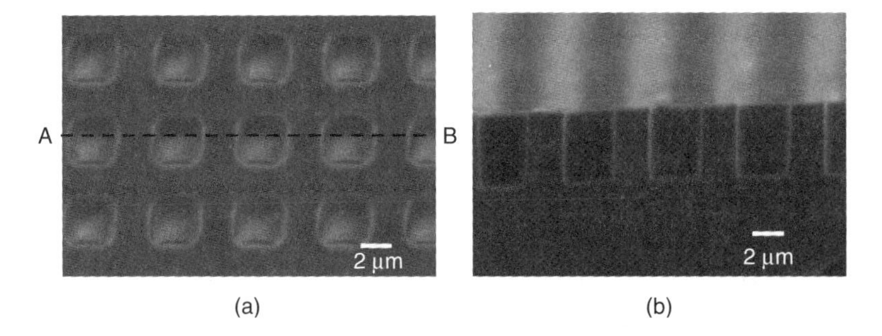

Figure 5.23 SEM image of samples fabricated by UV roller imprinting: (a) top view, (b) cross-sectional view (along line A–B in (a))

Figure 5.23 shows a SEM image of a resin pattern transferred onto a polyethylene terephthalate (PET) sheet; the image was obtained at a UV intensity of 50 mW/cm^2, slit width of 3 mm, and scanning stage speed of 67 mm/min. A cyclo-olefin polymer (COP) film was used for the mold. The film mold had dot patterns with a width of 3 μm, height of 5 μm, and pitch of 6 μm. After UV roller imprinting, hole patterns with a width of 3.4 μm and depth of 4.5 μm were formed.

High-throughput UV roller imprinting requires the stage to scan at high speeds [20, 21]. However, a fast scanning speed leads to a decrease in the UV exposure and filling time. A resistance heater and a temperature sensor are integrated with the roller shaft. Since the viscosity of the photocurable resin decreased at elevated temperature, the filling behavior of the resin was expected to improve by heating the roller. Figure 5.24 shows the relationship

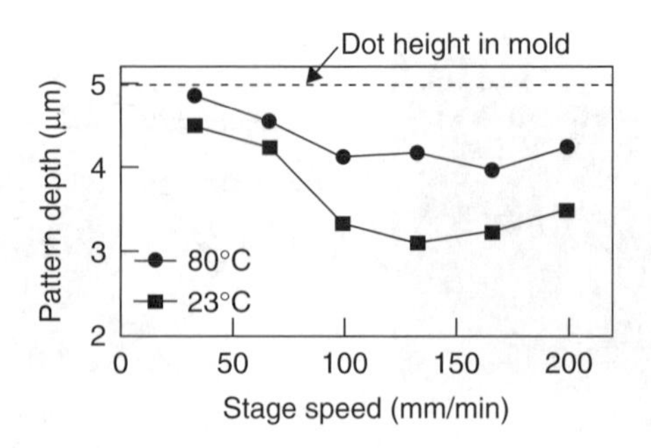

Figure 5.24 Relationship between pattern depth and stage speed at UV intensity of 32 mW/cm^2

between the depth of the transferred pattern and the stage speed at a UV intensity of 32 mW/cm^2 (slit width of 1 mm) and roller temperatures of 23 °C and 80 °C. The depth difference from the ideal value (5 μm) increased with the stage speed. However, the depth difference at 80 °C was smaller than that at 23 °C. This result indicates that heating the roller is effective in high-throughput UV roller imprinting.

5.3.2 UV Nanoimprint Process Tool

5.3.2.1 UV Nanoimprint Process

Figure 5.25 shows a process outline of a UV nanoimprint. Replication by UV nanoimprint is carried out using the following procedure. Firstly, a UV-curing resin is deposited or spin-coated on a substrate (a). Then, a quartz mold or UV transparency material mold as polymer film is approached and contacted on the substrate, and UV resin is filled into the mold by press force to the mold (b). After filling the resin in the mold, UV light is emitted through the mold and cures the resin (c). Replicated microstructures are complete on the substrate after releasing the mold from the substrate (d). Treatment of the release layer on the mold surface is necessary to separate the mold and UV resin easily and obtain a good replication. The UV nanoimprint process has the advantage of high-accuracy replication, because of the low stress and lack of heat deformation during the process, so that many kinds of application requiring sub-micrometer order patterning, such as patterned media, micro-optical devices, electronics devices, etc., could be realized using the UV nanoimprint process.

UV-imprinted pattern examples are shown in Figure 5.26. A 50 nm half pitch nanostructure is clearly replicated by the UV nanoimprint process.

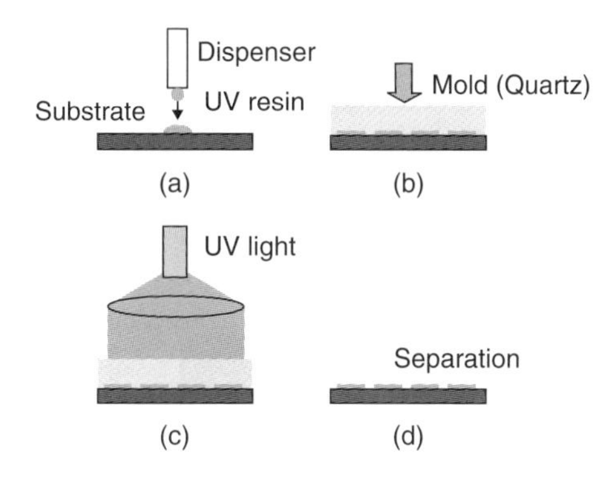

Figure 5.25 UV nanoimprint process

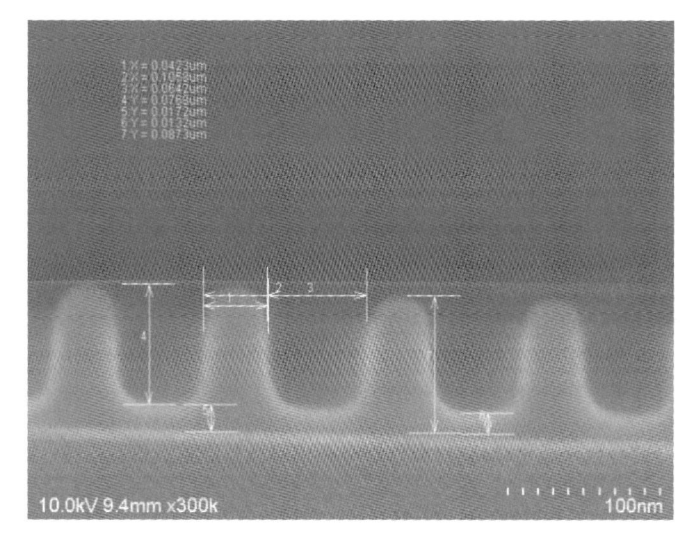

Figure 5.26 UV nanoimprinted pattern (50 nm half pitch)

5.3.2.2 Nanoimprint Process Tool

Several types of nanoimprint tool are available: a flat surface press type, a step and repeat type, and a roller press type as shown in Figure 5.27. It is important to select the tool type to meet the process requirements. The flat surface press type (a) is the most simple and standard type in the nanoimprint process. However, some technical issues in large-area replication have to be solved, which maintain the press force and illumination intensity uniformity

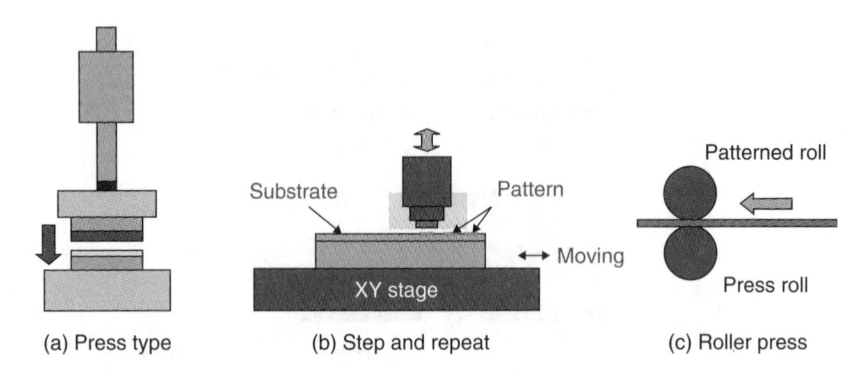

Figure 5.27 Nanoimprint process tool type: (a) press type, (b) step and repeat type, (c) roller type

in a large area, and the release procedure. The step and repeat type (b), which repeats the press type nanoimprint process onto the substrate in order, is a relatively reasonable way to form the nanostructure in a large area. However, the nanostructures formed at each step are independent of each other, hence the formation of a seamless nanostructure of large size by stitching at each step is basically not available. A roll press type (c), which utilizes a roller as a press mechanism, is the most attractive way to realize a large-sized microstructure because pressure uniformity in the line contact press region is easily maintained compared with the surface press type. Using the roll-type process, a roll-to-roll production system to replicate microstructures onto a polymer sheet could be possible.

(i) Press type
Figure 5.28 shows one example of a flat surface press-type nanoimprint process tool capable of 50 kN of press force and a step and repeat operation for a 100 mm square region with sub-micrometer positioning accuracy. The system consists of a motor-controlled press mechanism, UV LED light source with 365 nm peak wavelength, XY stage, and vacuum chamber to control the process conditions. A patterning example from the step and repeat operation is shown in Figure 5.29.

(ii) Roll-to-roll type
The system outline of a roll-to-roll-type nanoimprint process tool for continuous nanostructure replication is shown in Figure 5.30. The system consists of a UV resin coater to coat the resin with a predetermined

thickness of polymer sheet, a press roll, a mold roll, a UV light source, and sheet feed rollers to feed the sheet with constant tension. UV resin is filled into the mold pattern at a certain press force by the press roll and cured with UV light. The patterned sheet is automatically remolded after UV curing by sheet tension. Figure 5.31 shows examples of replicated patterns on a PET sheet of 100 μm thickness by the roll-to-roll nanoimprint process. Micro- and nanostructures are clearly replicated in the film.

5.3.2.3 Future Work

The UV nanoimprint process is a very attractive micro-/nanofabrication procedure. Many kinds of application could be promised using its process

Figure 5.28 Press type of UV nanoimprint tool

Figure 5.29 Patterning example of step and repeat nanoimprint process

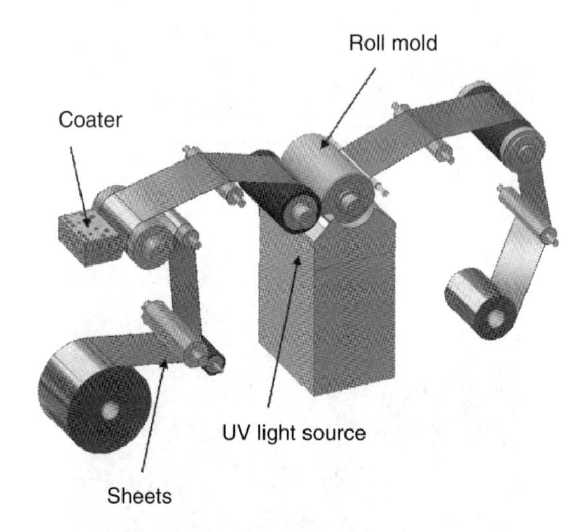

Figure 5.30 Roll-to-roll type of UV nanoimprint system

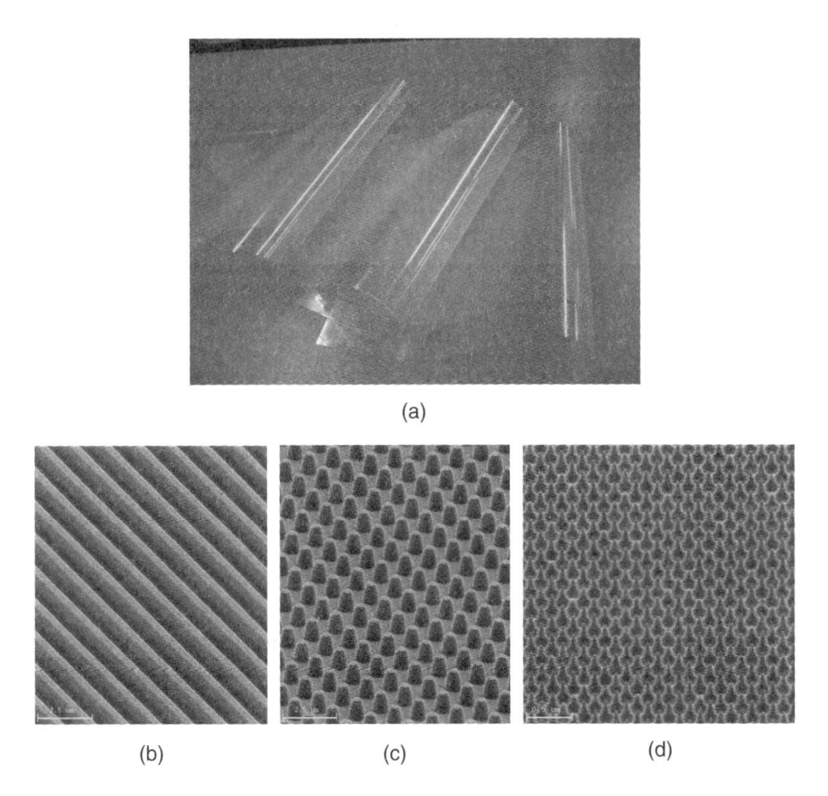

Figure 5.31 Pattern examples of UV roll-to-roll nanoimprinting: (a) patterned sheet, (b) 1 μm stripe pattern, (c) 1 μm dots pattern, (d) 200 nm pitch moth-eye structure

advantages, such as low stress and lack of heat cycle process. In order to accelerate the industrialization of the nanoimprint process, large-area and high-throughput nanoimprint processes and tools must be developed. A roll-type process is one reasonable solution to meet the demands of industrialization.

5.3.3 UV-photocurable Resin

Liquid photocurable resin is used as UV nanoimprint resin [22, 23]. Studies on the photocurable resin are being conducted along with the development of UV nanoimprint technology. Fluidity of the resin onto which the images are transferred influences the processing time greatly, because the nanoimprint technique is a molding process technology. In a UV nanoimprint, using low-viscosity photocurable resin can reduce the processing time and is best suited for mass production. Liquid photocurable resin is a type of resin that changes from liquid to solid with the activity of light. This change occurs by the polymerization reaction of monomer (oligomer) constituents in the resin. In this process, a photopolymerization initiator activates polymerization. The initiator absorbs energy efficiently from a specific wavelength of light from the optical source, and generates cure-initiation materials. Resin is classified according to the polymerization system into several types: radical polymerization type, ion polymerization type, ene-thiol type, etc. Among photocurable resin materials, the mainstream one is of photoradical polymerization type, but other types of material are increasingly being studied. Each type has its own material and curing properties, and it is desirable that the resin be chosen considering both the requirements of the application and the merits of the resin. The properties of resins with different types of curing mechanism and developmental examples will be described in the following.

5.3.3.1 Free Radical Polymerization

Free radical polymerization resin contains a monomer or oligomer with vinyl or (meta-) acryl, which enables radical polymerization, as well as an initiator of photo-initiated radical polymerization (Figure 5.32). There are two initiator types for photo-initiated radical polymerization: the cleavage type and the hydrogen abstraction type. The merit of using radical polymerization resin is its higher cure rate and the availability of a variety of materials. The necessary cure time is less than several seconds for a thickness range from several tens of nanometers to several tens of micrometers, although the exact timing depends on the light intensity and the thickness. Since most

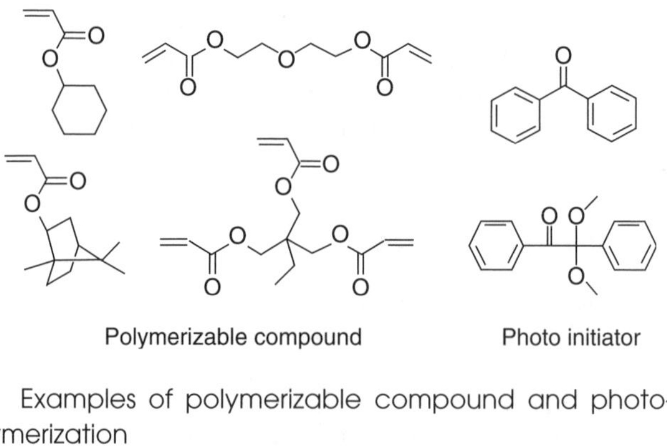

Polymerizable compound

Photo initiator

Figure 5.32 Examples of polymerizable compound and photo-initiator for radical polymerization

acryl vinyl monomers can be used for this type of resin, it is possible to control the solid-state properties of the resin relatively easily, and this type of resin is used in many applications. The drawbacks are cure inhibition due to oxygen, volume contraction during the cure [24], and relatively poor heat-resistance properties.

Radical polymerization resin is also widely used as a UV-NIL resin. Typical examples are the photocurable resin PAK-01 (Toyo Gosei [25]) and the low-viscosity resin (Molecular Imprints [26]). The former resin is of liquid form, and allows for spin coating. It has such merits as good coating properties, release properties, and transferring properties. In a report on the latter, the resin is made up from three kinds of acryl monomer – ethyleneglycol diacrylate, *t*-butyl acrylate, and (3-acryloxypropyl tris trimethylsiloxy) silane – and the liquid polymerization initiator DAROCURE1173. Selecting among the material constituents of the resin and optimizing their mixing ratio, the mechanical strength of the cured material is successfully improved. It is reported that the improved mechanical strength contributes to the suppression of transfer failures.

5.3.3.2 Cationic Polymerization

Cationic polymerization resin is a material containing a monomer or oligomer with cationic polymerization capability, such as epoxy or vinyl ether compounds, and a cationic photopolymerization initiator (Figure 5.33). The cationic photopolymerization initiator is mainly a photo-acid-generating agent: aromatic sulfonium salt or aromatic iodonium salt. Compared with

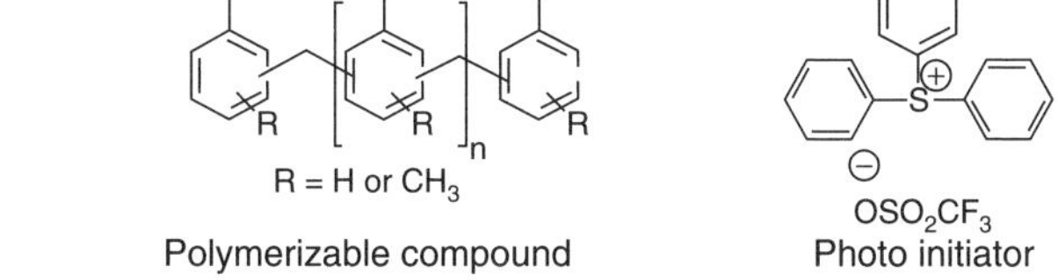

Figure 5.33 Examples of polymerizable compound and photo-initiator for cationic polymerization type

radical polymerization resin, this type of resin is less susceptible to cure inhibition due to oxygen. Volume contraction is small, and the heat-resisting properties are excellent. Its disadvantages are slow cure, a narrow range of material choice, susceptibility to temperature and humidity during the cure, and limited applications because of the acid that remains in the system.

Three examples of UV-NIL cationic polymerization resin will be described. First, acryl radical polymerization resin and epoxy cationic curable resin were made and compared [27]. The results of thermogravimetric analysis show the improved heat-resisting properties of the cationic polymerization resin. By using an organic–inorganic hybrid-type epoxy cationic curable resin, the temperature of 5% weight loss is increased from 270 °C for radical polymerization resin to 340 °C.

Ito *et al.* [28, 29] reported on the development of cationic curable resin using a vinyl ether compound. The vinyl ether compound is characterized by low viscosity compared with other types of monomer, and the photocurable resin using this material also shows low viscosity [30]. A transfer of 50 nm L/S was achieved using this resin. It was showed further that vinyl ether containing silicon is also applicable to UV-NIL if the materials are properly selected. Iyoshi *et al.* [31] reported on mold release using a hybrid resin of cationic and radical polymerization types. In this report, it is shown that the sensitivity is improved by optimizing the composition, and the mold removability is improved by suitably selecting the release agent.

5.3.3.3 Ene-thiol

Ene-thiol-type resin is optically cured by step reactions between a compound (ene) with two or more double bonds in the molecule and another compound with two or more thiols in the molecule. The reacting species are radical, but the cure inhibition by oxygen is less obvious compared with other known types of radical polymerization resin, as the thiyl radical that is produced is active and can react with inactive peroxy radicals. Another merit of step polymerization is that the material does not shrink too much. The drawbacks are that there is no variety of materials available and thiols are odorous, although less odorous thiols are being developed. This type of material has not yet been studied extensively as UV-NIL resin [32], but development is expected in the future.

5.3.3.4 Silicon-Containing Resin

We can expect a broader process window from the use of a multistep etching process. The multistep etching process uses a multilayer structure of resin materials. Figure 5.34 shows the process. There is a report, for example, where

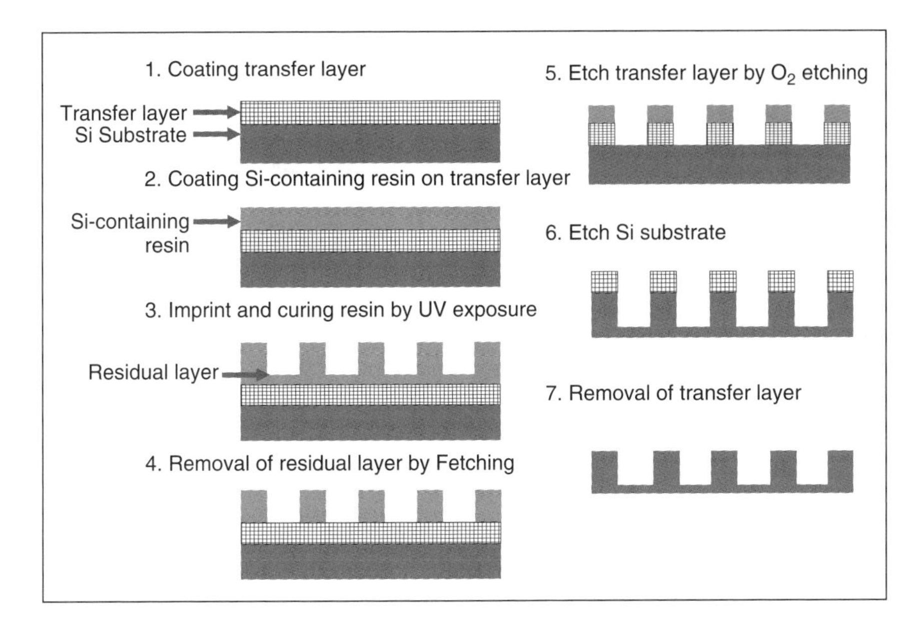

Figure 5.34 The multistep etching process using a bilayer structure of Si-containing resin and transfer layer

a silicon-containing material is used as the upper layer of the photocurable resin in the bilayer structure [24]. Silicon adds tolerability to oxygen dry etching. The tolerability to dry etching increases with the silicon content, and so the choice of monomer is important.

5.3.3.5 Fluorine-Containing Resin

When resin with high fluorine content is cured, its surface energy becomes small and it has excellent releasability. It is found that the contact angle of water on the cured surface of photocurable resin comprising a fluorine-containing monomer exceeds 90° [33]. Further, it is suggested that thermoplastic fluorine-containing polymer can be used as molding material for a UV nanoimprint (Figure 5.35).

5.3.3.6 Removable Resin

A difficult part of nanoimprinting is the removal of resin, which tends to adhere to the mold. Wet-type removal is almost impossible because

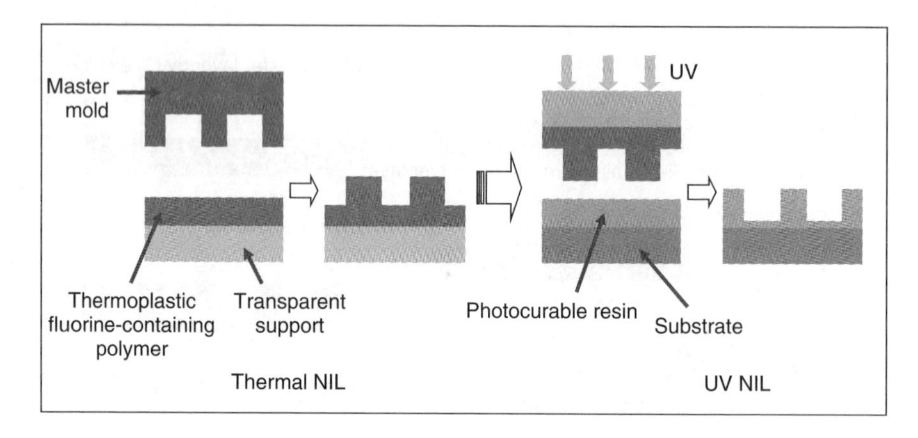

Figure 5.35 Replica method using fluorine-containing polymer.

a cross-linking agent is added to the photocurable resin to ensure the mechanical strength required in the release process. A possible solution for this problem would be to embed a structure in the polymerized resin that allows for chemical decomposition. For example, by emulating the design of the chemical amplification resist, it is proposed to use a monomer containing the *t*-butyl ester group and ketal group, which are dissociated in the presence of acid (Figure 5.36) [34].

There are many items that characterize resin. In addition to those discussed here, resistance to climate, refraction index, and other chemical and physical properties must also be improved. Improvement of material technology is indispensable for the progress of UV nanoimprint technology. Since many

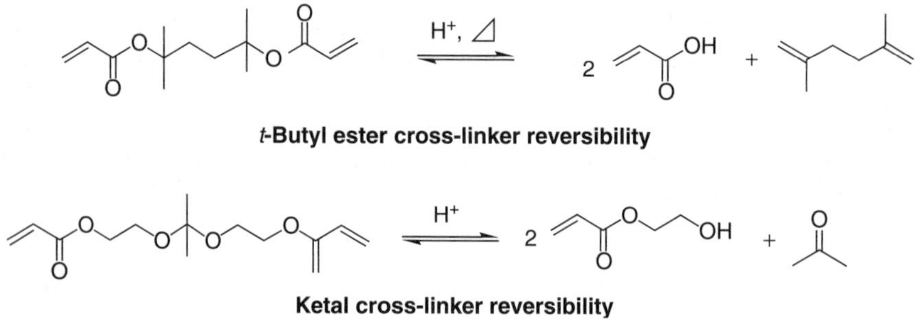

H^+, Δ

***t*-Butyl ester cross-linker reversibility**

H^+

Ketal cross-linker reversibility

Figure 5.36 Design of removable resin systems

institutions are now actively making an effort to improve resin, it is expected that superior resins will be available in the near future.

5.3.4 Fluorinated Polymers for UV-NIL

5.3.4.1 Introduction

UV nanoimprinting is now widely recognized as a highly productive and low-cost method to produce patterns with micro- and nanostructures. To apply this technology to actual mass production, it is essential to control the mold release properties.

In the usual case of UV nanoimprinting, a quartz mold with its surface coated with a release agent is used. Although release agents are very effective in decreasing the adhesion strength between the mold and the UV-curable resin, the release agent gradually degrades and the adhesion strength increases with repeated imprints. Strong adhesion will cause defects in the transferred pattern and may also damage the mold, so the mold must be treated frequently with the release agent. To ensure that the mold is uniformly treated, the treatment process needs many troublesome steps such as removing the remaining release agent, recoating the release agent, and drying.

In order to improve the productivity of UV nanoimprinting, we have focused on developing materials for a release agent-free nanoimprint process. Our strategy was to use fluorinated polymers as the mold material or the UV-curable resin [35]. The advantage of fluorinated polymers is that they themselves have good mold release properties so there is no need to use any release agent. In this section, the two materials we have developed, "F-template" and "NIF," will be introduced.

5.3.4.2 F-template: Perfluorinated Polymer Mold Material for UV Nanoimprinting

F-template is composed of a thermoplastic resin coated on a substrate [35]. The nanoimprint process using F-template is shown in Figure 5.37. First, a replicated mold is fabricated from the F-template by thermal nanoimprinting. Then, the replicated mold is used to produce nanostructures by UV nanoimprinting.

The thermoplastic resin used in the F-template is an amorphous perfluorinated resin which has many characteristics suitable for the nanoimprint process. It has extremely low surface energy, so no release agents are needed in either the thermal or the UV nanoimprint process. It has high resolution, and nanoimprinting of patterns with dimensions as small as 50 nm has

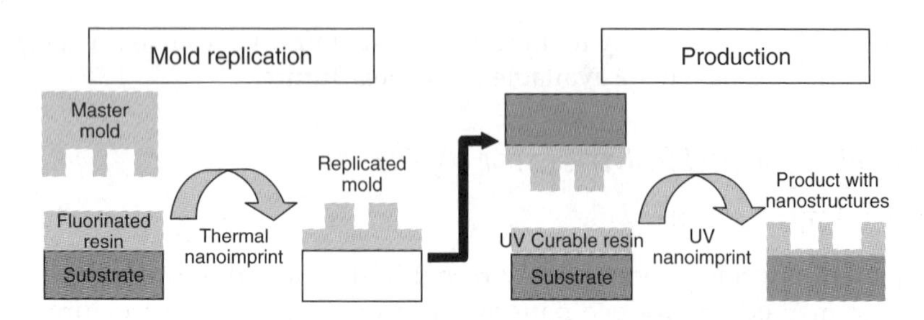

Figure 5.37 Nanoimprint process using F-template

been confirmed [36]. It has very high transparency and stability to UV light (wavelength >200 nm), so various light sources can be selected in the UV nanoimprint process. It is also inert to most chemicals and solvents.

The substrate material of the F-template can be selected from various materials, such as quartz glass, borosilicate glass, silicon, aluminum, PET, and polycarbonate. The form of the substrate may be a rigid sheet, a flexible film, or a non-planar structure such as a cylinder. Flexible film-type and cylinder-type F-templates can be used to fabricate replicated molds for a roll-to-roll nanoimprint process [35].

Durability of the replicated mold – in other words, the number of times the replicated mold can be used in the UV nanoimprint process – is an important factor in designing the nanoimprint process. In order to examine the durability, we fabricated replicated molds with line and space patterns, pillar patterns, and hole patterns with dimensions of 0.5 μm to 2 μm and repeated UV nanoimprint. A commercially available UV adhesive was used as the UV-curable resin. No significant change was observed in the pattern dimensions, even after repeating the UV nanoimprint 500 times (Figure 5.38). There was only a slight decrease in the contact angle of water (110° to 108°).

The key point in the durability tests was to apply pressure in the UV nanoimprint process. When there was no pressure applied, the UV adhesive started to adhere and remain on the patterns of the replicated mold after repeating the UV nanoimprint about 100 times. The amount of UV adhesive remaining on the mold gradually increased until the pattern was covered completely with the UV adhesive. This phenomenon was significant near the border between the area with patterns and the area without patterns. When a pressure of about 1 MPa was applied, no UV resin remained on the replicated mold. The mechanism of adhesion is not yet clear, but the curing shrinkage of the UV adhesive seems to have an important role. When there is no restriction, the UV adhesive shrinks about 10% while it is cured. We suppose that the UV adhesive remains on the mold by physical anchor

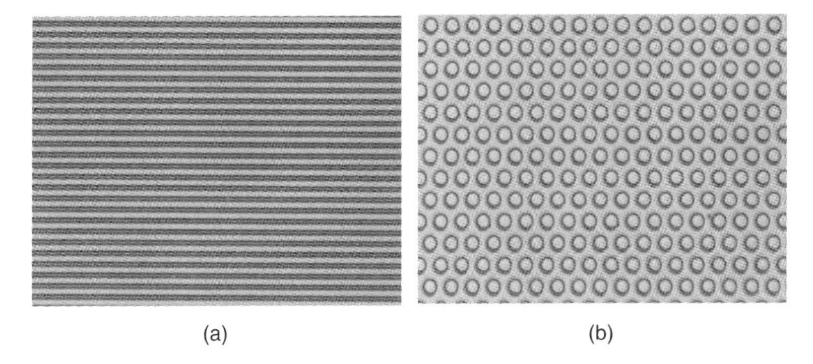

Figure 5.38 Confocal laser microscope images of nanostructures fabricated at the 500th UV nanoimprint. (a) Half pitch 0.5 μm line and space pattern, (b) 1 μm pillar pattern

effect, caused by this curing shrinkage. When pressure is applied in the UV nanoimprint process, the shrinkage of UV-curable resins occurs mainly in the residual layer and the pattern dimensions are maintained. Thus, the physical anchor effect is minimized and the UV adhesive is easily detached from the mold.

The F-template is used to fabricate many replicated molds from one master mold, but it can also be used to fabricate large-sized molds by the thermal step and stamp imprint lithography method (SSIL) [36]. In SSIL, a small master mold and a larger-sized F-template are used. The thermal nanoimprint is repeated until the nanostructure of the mold is formed on all the required areas of the F-template, and a large-size replicated mold is obtained.

The F-template can be applied to nanopatterning methods other than the UV nanoimprint, such as nanocasting lithography [37] and electroforming. In the electroforming process, a thin layer of metal is sputtered on the F-template with nanostructures to form a conductive layer and then nickel is electrodeposited. After the F-template is removed, a nickel mold with nanostructures on its surface is obtained. In order to completely remove the F-template, the perfluorinated resin is dissolved with fluorinated solvents such as ASAHIKLIN AC-6000, a product of Asahi Glass Co., Ltd.

5.3.4.3 NIF: UV-curable Resin with Fluorinated Components

NIF is a composition of UV-curable acrylate compounds and fluorinated compounds [35]. It has excellent mold release properties with quartz, silicon, or nickel molds, even without any release agents. NIF-A-1, a standard grade

Table 5.1 Properties of UV-curable resin ''NIF-A-1''

	Units	NIF-A-1
Viscosity (25 °C)	mPa s	16
Sensitivity (@365 nm, thickness 1.5 μm)	mJ/cm^2	>441
Tensile elasticity	MPa	2100
Contact angle of water	°	93
Curing shrinkage	%	9
Transmittance (>362 nm, thickness 15 μm)	%	>80
Refractive index (@589 nm, 23 °C)	–	1.49
10% weight loss temperature	°C	220
Dry-etching resistance (CF_4)	vs. PMMA	0.5

of NIF, has various characteristics such as low viscosity, high sensitivity, low surface energy, high transparency, and high dry-etching resistance, as shown in Table 5.1. NIF has a very high resolution, and imprinting of line and space patterns with half pitch of 21 nm has been confirmed. NIF can be used as an etching resist [38] or as a permanent component for optical devices.

In order to evaluate the mold release properties, NIF-A-1 was cured between a bare quartz plate and a quartz plate with a primer coating, and the force needed for demolding was measured. The demolding force of NIF-A-1 was about one-fifth that of a non-fluorine-containing UV-curable resin, and NIF-A-1 could easily be detached from the bare quartz plate. When a quartz plate treated with a release agent was used instead of the bare plate, there was almost no difference in the demolding force between NIF-A-1 and non-fluorine-containing UV-curable resin. However, it has been reported that when NIF-A-1 is used as the UV-curable resin, the surface energy of the release agent-treated mold does not change even after repeating the UV nanoimprint 500 times [38]. The surface energy of the mold gradually increased when a non-fluorine-containing UV-curable resin was used. Thus, the mold release properties can be improved by using NIF as the UV-curable resin, either with or without release agents on the mold.

NIF has excellent release properties, but as a consequence, adhesion of NIF to inorganic substrates (glass, silicon, etc.) is weak. Surface treatment of the inorganic substrate by methacrylate silane coupling agent is effective in improving adhesion. Organic coatings are also effective [39].

As mentioned in Section 5.3.4.2, the pattern dimensions of UV-curable resin are maintained by applying pressure in the UV nanoimprint process. The curing shrinkage of NIF-A-1 is about 9%, but by optimizing UV nanoimprint conditions, the shrinkage of the pattern dimensions can be suppressed to less than 1%.

The properties of NIF can be controlled by its composition. For example, we have developed NIF with dry-etching resistance as high as a KrF resist, NIF with a refractive index of 1.41 to 1.57, NIF with higher light resistance, NIF with higher sensitivity, NIF with higher viscosity or lower viscosity. The required properties are different for each application, and we are continuing to develop new compositions of NIF.

5.3.4.4 Mold Replication Using NIF

NIF, with its excellent mold release properties, can also be used as a material for mold replication. In the case of NIF, replicated molds are fabricated by UV nanoimprint. Then, the replicated mold is used to produce nanostructures. The F-template and NIF are compared in Table 5.2.

NIF has a higher surface energy compared to the F-template, which can be estimated by the contact angle of water. In order to use a NIF mold many times, it is desirable to lower its surface energy to improve the mold release properties. We examined the effect of the curing conditions and found that the surface energy of the master mold had a large influence on the surface energy of the cured NIF [40]. When NIF was cured in contact with a hydrophilic mold, the contact angle of water on the cured NIF was around 93°. The UV nanoimprint was repeated with this NIF mold, and the contact angle of water decreased rapidly with several imprints. When NIF was cured in contact with a fluorinated mold, the contact angle of water increased to 105° and its durability against UV imprint was improved [40].

Recently we have developed the "anti-sticking cure process" (ACP), a new process to improve the mold release properties of NIF molds. In ACP, first a

Table 5.2 Comparison of F-template and NIF as materials for mold replication

	F-template	NIF
Mold replication method	Thermal nanoimprint	UV nanoimprint
Production method	UV nanoimprint	
Material	Perfluorinated thermoplastic polymer	UV-curable resin containing fluorinated compounds
Substrate	Glass, polymer film, silicon, etc.	
Contact angle of water	>110°	>90° (>100° after improving the curing process)

NIF mold is fabricated by UV nanoimprint and after it is detached from the master mold, the NIF mold is irradiated with a strong UV light (wavelength 300–400 nm). When ACP is applied, mold release becomes much easier in the UV nanoimprint using NIF molds. The mechanism is not yet clear, but the contact angle of water on the NIF mold increases with ACP, so fluorinated components seem to be diffusing to the surface. We also suppose that active functional groups remaining on the surface of NIF are eliminated by ACP. Details are now under investigation.

One advantage of using NIF for mold replication is that there is no thermal process and only UV nanoimprint tools are needed. UV nanoimprint takes much less time than thermal nanoimprint, and many replicated molds can be fabricated in a short time. There are applications in which there is a high risk of accumulating defects when one mold is used many times. For such applications, NIF can be used as a disposable mold.

5.3.4.5 Summary

We have introduced the F-template and NIF as materials for a release agent-free UV nanoimprint process. The F-template, with its high stability, is suitable for making highly durable replicated molds. Also, large-area molds can be fabricated from the F-template by the SSIL process. NIF has excellent mold release properties and can be applied to various applications such as dry-etching resists, permanent components, and disposable molds. We believe that the productivity of the UV nanoimprint process will be dramatically improved by a release agent-free UV nanoimprint process.

5.3.5 Cationic Curable Resins for UV-NIL

5.3.5.1 Introduction

A photocurable system is acknowledged to be an environmentally friendly technology, as it is a faster, more efficient and effective energy-saving process than the heat-curable process, having widespread applications, especially in coating industries [41]. These features are being exploited for nanoimprint technology to become an alternative to conventional photolithography [42–44].

Cationic curing is a polymerization reaction induced by cations or strong acids. In a UV cationic system, the cation generates UV exposure with decomposition of a photo-initiator or photo-acid generator (PAG). The monomers that undergo cationic curing are cyclic ethers such as epoxides [45], oxetanes [46], and so on, and vinyl ethers [47]. Figure 5.39 shows typical types of

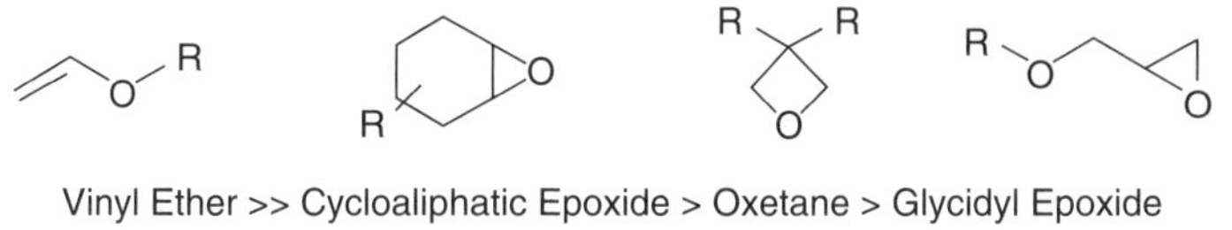

Vinyl Ether >> Cycloaliphatic Epoxide > Oxetane > Glycidyl Epoxide

Figure 5.39 Basic molecular structures of cationic curable resins and the order of cationic reactivity

cationic curable monomers along with their rough order of UV curing rate. Two types of epoxide – cycloaliphatic epoxide and glycidyl epoxide – are illustrated here. While the glycidyl epoxide is widely used in industrial and household adhesives and must be familiar to most people, it is the cycloaliphatic epoxide (one of Daicel' s core products) that plays the leading roll in a UV cationic curing system due to its commercial availability as well as quick curability.

A UV-curable system has two main categories of reaction type: radical curing and cationic curing. Each type has its own features. Here, the cationic curing system is introduced in contrast to the UV radical curing system.

5.3.5.2 Characteristics of UV Cationic Curable Resins

UV radical curing resin consists mainly of UV radical initiator, and acrylic monomers and oligomers. Radical curing reactions run very fast. No more than 1 s of UV exposure substantially completes the curing. A wide variety of acrylic compounds are commercially available, which allows formulations to be established for various nanoimprint applications. For this reason, acrylic compounds have been studied widely in the UV nanoimprint. One concern with the radical curing system is oxygen inhibition. Oxygen has a detrimental effect on the cure response of free radical systems, especially in thin-film coatings. For films thinner than several hundred nanometers, oxygen inhibition may be unavoidable unless the reaction runs under reduced pressure. For the nanoimprint process, however, pressing the mold to the resin may have the effect of avoiding oxygen inhibition.

UV cationic curing resins consist mainly of PAG and multifunctional cationic monomers and oligomers. Each type of cationic curable monomer has its own features. Table 5.3 represents the characteristics of typical cationic monomers in contrast to the typical characteristics of acrylates. Among the cationic monomer types, vinyl ether behaves quite differently from the rest of the monomer types. Vinyl ether reacts as fast as the radical curing, and at the same time shows large shrinkage rates comparable to those

Table 5.3 Comparative properties among UV curable monomers

	Radical curing type	Cation curing type			
	(Metha) Acryl	Glycidyl epoxide	Aliphatic epoxide	Oxcetane	Vinyl ether
Reactivity	Excellent	Poor	Good	Good	Excellent
Dark reaction	Not occur	Available	Available	Available	Available
Heat resistance	Good	Good	Excellent	Good	Poor
Shrinkage	>10%	<5%	<5%	<5%	>10%
O_2 inhibition	Affected	Unaffected	Unaffected	Unaffected	Unaffected
Viscosity range	Low to High	High	High	Low to High	Low to High

for acrylate. The advantages and disadvantages of the UV cationic curable process – except for vinyl ether – are summarized as follows.

Advantages

(1) Curing shrinkage far lower than for the radical curing process.
(2) Cationic reaction system has no curing inhibition by oxygen.
(3) Curing reaction may continue after UV exposure (dark reaction).

Disadvantages

(1) The presence of moisture or alkaline may in some cases slows the curing reaction.
(2) Cationic curing rates of cyclic ethers are supposed to be lower than the radical curing rate. However, this depends greatly on the photo-initiators and monomers, as discussed later.

5.3.5.3 Curing Shrinkage

One advantage of introducing the cationic curing system in the nanoimprint may be the lower curing shrinkage than for the radical curing system. A detailed study [48] by R.F. Brandy *et al.* explains that the difference in shrinkage is attributable to their polymerization styles. The radical polymerization of acrylates takes additional polymerization, and the cationic polymerization of epoxides and oxetanes takes ring-opening polymerization, except

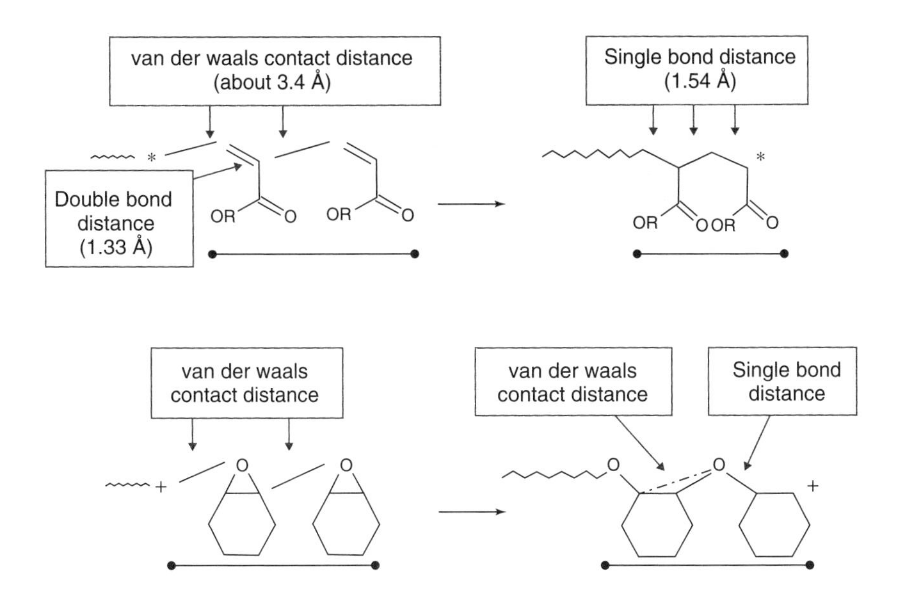

Figure 5.40 Factors that contribute to volume change during polymerization

for the cationic polymerization of vinyl ethers which falls into additional polymerization. The mechanism of curing shrinkage is shown in Figure 5.40. The distance between monomers changes from the van der Waals contact distance (approximately 3.41 Å) to the covalent distance (1.54 Å) on polymerization reaction of the monomers. The upper figure indicates the additional polymerization of an acrylate. The lower figure shows the ring-opening polymerization of cycloaliphatic epoxides. The distance between the monomers becomes short with ring-opening polymerization, as with additional polymerization. However, the ring-opening polymerization cleaves one C–O covalent bond on the epoxy group, appearing as an ether bond between monomers, and resulting in the distance between the two atoms being the van der Waals contact distance. That is why the total shrinkage of the ring-opening polymerization becomes smaller than for additional polymerization.

Figure 5.41 shows one example revealing the difference in curing shrinkage between the radical and the cationic curing system. A 20 μm layer was formed on a strip of polyimide film and then UV irradiated for curing. The degree of curvature deformation of the polyimide ribbon was observed. The radical type having 10–12% shrinkage rate largely curved toward the applied resin side, while the cationic type having 3–5% shrinkage rate exhibited a slight curve toward the opposite side of the resin. Although it is still not understood

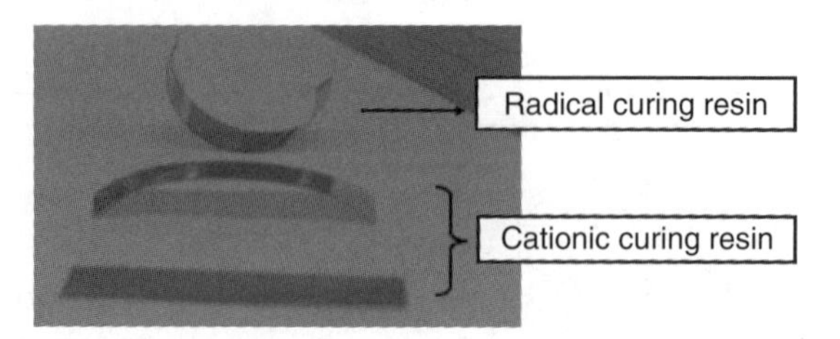

Figure 5.41 Comparison of shrinkage between cationic curing resin and radical curing resin

how the curing shrinkage of the resin affects the resolution of the pattern transfer, the low shrinkage rate definitely errs on the right side.

5.3.5.4 Curing Rate

Although the cationic curing rates of cyclic ethers are supposed to be lower than the radical curing rate, this is largely dependent upon photoinitiators, monomer types, and their combinations to be formulated. With the knowledge and experience Daicel has accumulated, it is not hard to achieve cationic formulations that cure in less than a few seconds. Another technique for getting quick and firm curing in the cationic resin is to utilize dark reaction, a unique characteristic of the cationic curing system, as the polymerization continues after UV irradiation has stopped. Post-exposure bake (PEB) accelerates the dark reaction. So, small doses of irradiation energy followed by heating (PEB) could accomplish sufficient curing in a short period of time. This method is used in chemically amplified resist [49].

One more method for the enhancement of cationic reaction is to formulate epoxides in conjunction with vinyl ethers. Figure 5.42 shows a real-time infrared spectrum (IR) chart of the two cationic formulations. One is for cyclohexene oxide with 3 parts of initiator Irg250 (iodonium(4-methylphenyl)[4-(2-methylpropyl)phenyl]hexafluorophosphate) from Ciba-Geigy and 50 mW/cm^2 intensity of UV irradiation (Figure 5.42(a)). The other is the result of a mixture of 50wt% cyclohexene oxide and 50wt% cyclohexyl vinyl ether with 3 parts of Irg250 and 50 mW/cm^2 intensity of UV irradiation (Figure 5.42(b)). Observing the absorption peak around 1100 cm^{-1} corresponding to the ether group resulting from the ring-opening reaction of the epoxy group, the rising speed of the peak in Figure 5.42(b) is much faster than that in Figure 5.42(a). This means that the vinyl ether has the effect of enhancing the ring-opening cationic curing reaction of the epoxide when the two substances coexist.

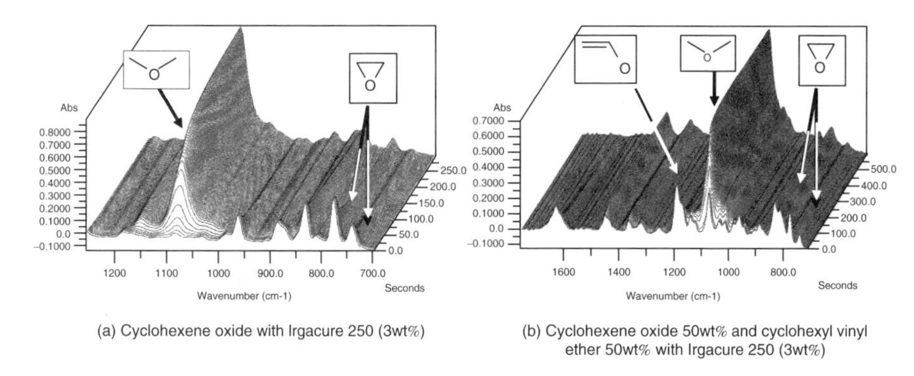

(a) Cyclohexene oxide with Irgacure 250 (3wt%)

(b) Cyclohexene oxide 50wt% and cyclohexyl vinyl ether 50wt% with Irgacure 250 (3wt%)

Figure 5.42 Effect of vinyl ether enhancing cationic reaction of epoxide

5.3.5.5 Demolding

An epoxy is generally thought to give better contact with substrates such as silicon wafer or a piece of glass than acrylate. This is one of the advantages over acrylate that the epoxy enjoys. One example is shown in Figure 5.43, expressing the difference in cross-tension strength between the radical and the cationic type.

From the very nature of the nanoimprint process where the template must get in touch with the resist or imprint resin, releasing the template from the resin after curing is the most troublesome task throughout the process. A popular technique is to apply a release agent on the template surface for easy demolding. The release agent generally used in the nanoimprint may be a fluorine compound having silanol groups in the molecular chain to make firm connection with hydroxyl groups present in the template surface by silane-coupling reaction. The treatment of the template with the release agent is found effective for easy demolding, but is also found insufficient in respect of its durability for repeated use.

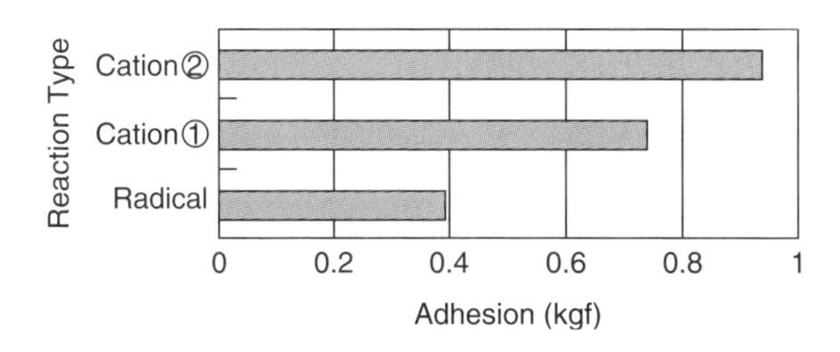

Figure 5.43 Cross-tension strength between glass plate and silicon substrate

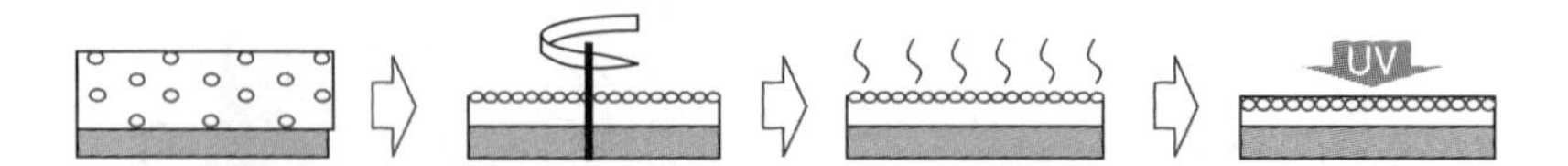

Figure 5.44 Conceptual diagram of the bleeding effect

To ease some part of this problem we have reviewed additives that may work for not only demolding, but also contact with the substrate. The additives are a kind of internal release agent, basically fluorine compound or silicone compound. Some of the additives, we have found, tend to migrate to the surface of the resin layer during pre-baking after spin coating, working to reduce the demolding force [50, 51]. This concept is called the "bleeding effect," and is illustrated in Figure 5.44. Two experiments were conducted to reveal the bleeding effect. One is to measure the separation forces or cross-tension strength between a piece of glass plate and a silicon wafer. The other is to analyze the cured resin surface by X-ray photoelectron spectroscopy (XPS) and measure the water contact angle.

The test method for cross-tension strength is as follows.

(1) Put a few drops of resin solution on a silicon wafer substrate of dimension 0.5 mm × 20 mm × 36 mm.
(2) The resin layer is formed by spin coating.
(3) Vaporize the solvent by pre-baking.
(4) Put a glass plate of dimension 1.2 mm × 20 mm × 36 mm criss-crossed on the silicon wafer substrate.
(5) Set the test piece in the nanoimprinter.
(6) Apply pressure and UV radiation for curing.

The tension strength was then measured.

Test pieces for surface analysis were prepared in the following manner.

(1) Put a few drops of resin solution on a silicon wafer substrate of dimension 0.5 mm × 20 mm × 20 mm.
(2) The resin layer is formed by spin coating.
(3) Vaporize the solvent by pre-baking.
(4) Apply UV radiation for curing.
(5) The test pieces undergo XPS analysis and measurement of water contact angle.

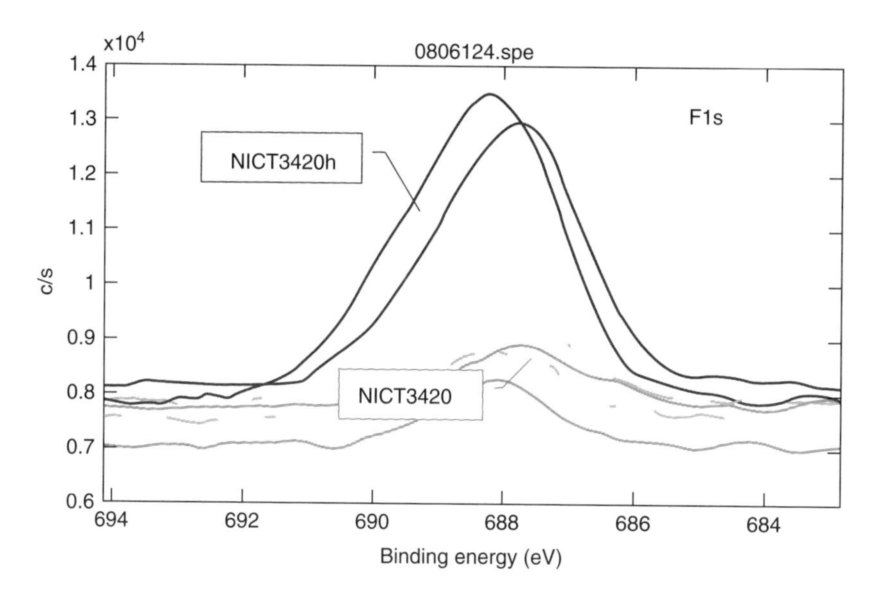

Figure 5.45 Fluorine concentration on cured resin surface

Figure 5.45 shows the concentration of fluorine atoms on the cured resin surface. NICT3420h containing 0.1% of additive "h" shows plenty of fluorine atoms detected on the resin surface, which indicates the bleeding effect. Here, "h" is a symbol bestowed on the fluorine compound that we found effective throughout the screening test. NICT3420 is the cationic curing-type resin we developed for nanoimprint and contains no additive. NICT3420, in contrast, had only a trace of fluorine atoms detected on the resin surface. These fluorine atoms of NICT3420 are assumed to originate from the photo-acid generator, as well as the fluorine compound that was used for the experiment.

Figure 5.46 shows the outcome of the bleeding effect exerted on the demolding force and water contact angle of the resin surface. In the resin containing 0.1 parts of the additive "h," the demolding force decreased by nearly 40% and the water contact angle increased by approximately 25% to 80° compared to the resin without additive. As at least a 90° water contact angle is supposed to be desirable for smooth demolding, an 80° water contact angle falls far short for the purpose, but some expectations could be built on the presence of a release agent on the resin surface helping the release treatment of the mold to survive longer than it otherwise could.

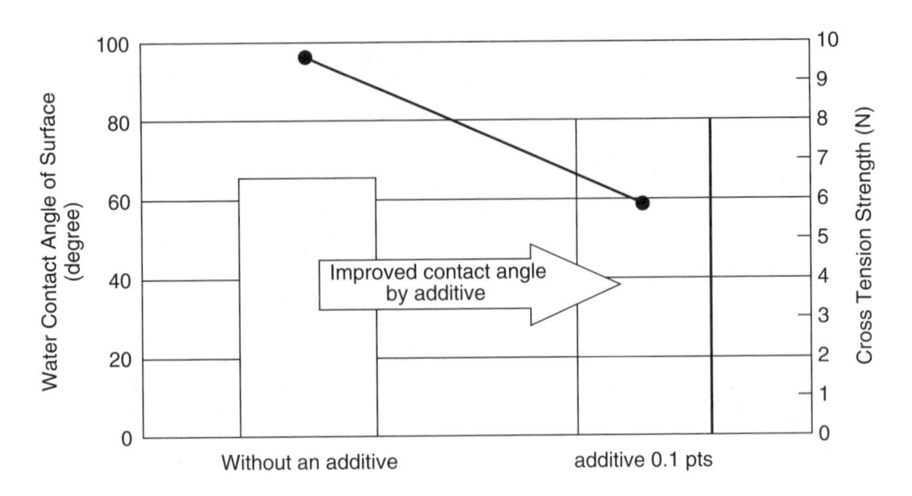

Figure 5.46 Improvement of template release force by additive

5.3.5.6 Transfarability

Figure 5.47 shows SEM images of a UV nanoimprint using cationic curing resin, imprinted under MPa imprint pressure and at room temperature. The line and space patterns are clearly transferred with sharp edges.

5.3.5.7 Activities of Daicel Corporation

Daicel Corporation has globally pioneered the excellence of cycloaliphatic epoxides by developing, manufacturing, and marketing a variety of their derivatives, and now is the single supplier of cycloaliphatic epoxides in the

Figure 5.47 SEM images of UV nanoimprint using cationic curing system

world. The cycloaliphatic epoxides, unlike glycidyl epoxides, have the edge in terms of environmental impact because they do not contain halogens like chlorine and bromine, a source of highly toxic dioxine if improperly incinerated. This comes from the way in which they are manufactured. While glycidyl epoxides are manufactured by epichlorohydrin, cycloaliphatic epoxides are made by direct oxidization or epoxidation of cyclohexenes with peracetic acid, one of Daicel's backbone materials.

Cycloaliphatic epoxide is good at cationic reactions to give a cured body with high glass transition temperature, heat resistance, and anti-yellowing properties, leading to wide industrial applications such as LED encapsulation. Daicel is directing all its cationic curing technology accumulated from the development of cycloaliphatic epoxides and derivatives toward designing and evaluating UV nanoimprint resins.

Daicel's typical UV-NIL resins are shown in Table 5.4. Each resin is categorized into two curing types: cationic curing type (NICT), radical curing type (NIAC). Both curing types have unique properties. NICT has good performance with PDMS (poly dimethyl siloxane) mold, and NIAC is suitable for lift-off process of nanoimprint lithography, because it can solve into some solvents after curing.

Daicel's UV-NIL resins have the features described below.

(1) Quick curing has been achieved even for extremely thin films, less than 100 nm in thickness.
(2) A wealth of line-ups are available that make it possible to obtain film thicknesses from 50 nm to several tens of micrometers.
(3) Resin viscosities can be designed from a few millipascal seconds to several thousand millipascal seconds as required for applications. Other properties like refractivity, transparency, and mechanical and thermal properties can be tailored to customers' needs. Samples featuring improved demoldability by the bleeding effect are also available.

5.3.5.8 Summary

We have developed cationic curable resins appropriate for UV nanoimprinting. Cationic curable resins using epoxide and oxetanes, which react in ring-opening polymerization style, indicated low shrinkage and good adhesion to the substrate. Vinyl ether showed acceleration of the whole cationic curing system. As a solution for improving the demolding, some additives were found effective for both demolding and good contact with the substrate. So, cationic curable resins were expected for UV nanoimprinting with high resolution and high throughput.

Table 5.4 Typical grades of UV-Nanoimprint resins

GRADES TYPE	UNIT	**NICT825** CATION	**NICT109** CATION	**NIAC705** RADICAL
VISCOSITY[1]	mPas/25°C	380	550	30
NON-VOLATILE CONTENT	%	100	60	100
SHRINKAGE[2]	%	4.4	-	7.3
REFRACTIVE INDEX[3]	-	1.51	-	1.53
PROPERTIES	-	Low Shrinkage Good Cotact to Substrate	Uniform Laye Multi-Patterning	Solvents Solubility

[1]E-type viscometer before dilution/NIAC109 contains some solvent without dilution.
[2]Calculated from liquid density and cured solid density/difference in density before and after curing.
[3]Abbe's refractometer.

5.3.6 Molding Agents for Nanoimprinting

5.3.6.1 Silane Coupling Agents

Silane coupling agents are silicon compounds represented by Y_nSiX_{4-n} ($n = 1, 2, 3$). Y represents either non-reactive groups including alkyl and phenyl groups or reactive groups having vinyl, amino, or epoxy groups at the terminal of the alkylene chain. X denotes hydrolyzable groups including halogen, methoxy, ethoxy, acetoxy, or isocyanato groups that react easily with water, hydroxyl groups, or adsorbed water on the substrate surface. If these hydrolyzable groups react with water, they change to hydroxyl groups, which then bind to hydroxyl groups on the surface of the substrate through condensation reaction. Silane coupling agents are widely used when composite materials consisting of organic and inorganic substances such as GFRP (glass fiber-reinforced plastics) are produced as a mediator between these substances. If Y is a non-reactive group such as alkyl or phenyl group, the coupling agents make the modified substrate surface adhesion- and wear-preventive, gloss-keeping, water-repellent, lubricative, non-staining, and moldable. When the agents have reactive groups like vinyl and amino groups, they are used to organize the substrate surface. Depending on the number of Ys, the agents are used as coupling agents ($n = 1$), materials for siloxane polymers ($n = 2$), and polymer blocking agents ($n = 3$) (end-capping agents for blocking both terminals of siloxane polymers) [52].

Compounds with fluorocarbon chains are water-repellent and oil-repellent as represented by poly(tetrafluoroethylene) (PTFE) and show such peculiar properties as non-flammability, high lubricability, chemical inactivity, low toxicity, etc.

Silane coupling agents prepared by introducing a straight fluorocarbon chain into the aforementioned Y of silane coupling agents greatly improve the substrate surface. Thus, when compared with PTFE surface, the surface modified with fluorinated coupling agents acquires a free energy as low as or even lower than that of PTFE surface, shows a higher lubricability and moldability, and becomes oil-repellent as well as water-repellent.

5.3.6.2 Mechanism of Surface Modification Using Silane Coupling Agents

Figure 5.48 shows the mechanism of reaction of common silane coupling agents with substrate surface. The addition of water to silane coupling agents causes them to undergo hydrolysis and then condensation to form oligomers composed of several coupling agent monomers. The oligomers formed make hydrogen bonds with hydroxyl groups or adsorbed water molecules on the substrate surface. Heat treatment is performed at 100–150 °C for several

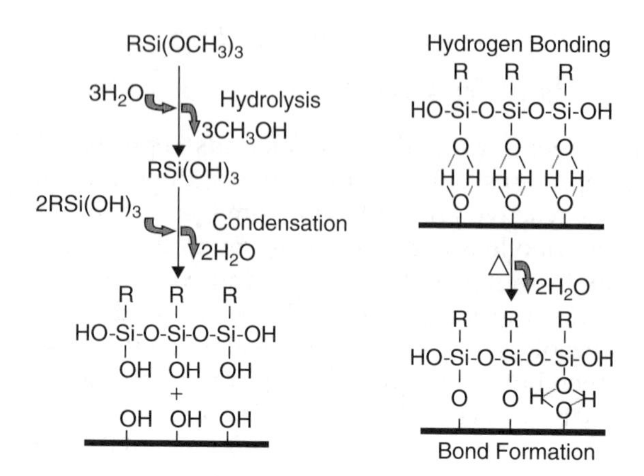

Figure 5.48 Reaction and binding mechanism of alkoxysilanes

tens of minutes to strengthen the binding. This treatment facilitates the dehydration process, causing hydrogen bonding to convert into covalent bonding. Although a silane coupling agent-modified substrate surface is relatively durable in dry conditions, the Si–O substrate bond formed undergoes hydrolysis in the presence of a small amount of water or moisture to revert to Si–OH and HO–substrate. That is, the silane coupling agent layer separates from the substrate surface and the characteristic properties of the modified surface are lost.

Fluorinated silane coupling agents ($RfCH_2CH_2Si(OCH_3)_3$, Rf = perfluoroalkyl chain (e.g. $F(CF_2)_8$-) or poly(fluoroalkyl ether) chain) have so far been used as silane coupling agents for nanoimprinting. Since the fluoroalkyl chain or poly(fluoroalkyl ether) chain attached to the terminal of the silane coupling agent is water- and oil-repellent, the substrate surface modified with the fluorinated coupling agent is hardly hydrolyzable and highly moldable because of its low surface free energy.

The thermostability of the substrate surface modified with $F(CF_2)_8CH_2CH_2Si(OCH_3)_3$ is nevertheless at the same level as that of poly(tetrafluoroethylene) (common thermostable temperature: 260 °C). Considering the molding frequency, the modified surface is not usable at the upper limit of thermostable temperature, and is used at around 150 °C. Those molding agents with poly(fluoroalkyl ether) chain which are commercially available and widely used are also thermally non-durable.

This section describes the synthesis, surface modifiability, and application to nanoimprinting of novel silane coupling agents, $Rf(C_6H_4)_2C_2H_4Si(OCH_3)_3$ and $Rf(C_6H_4)_2CH_2CH_2CH_2Si(OCH_3)_3$, obtained by introducing a biphenyl

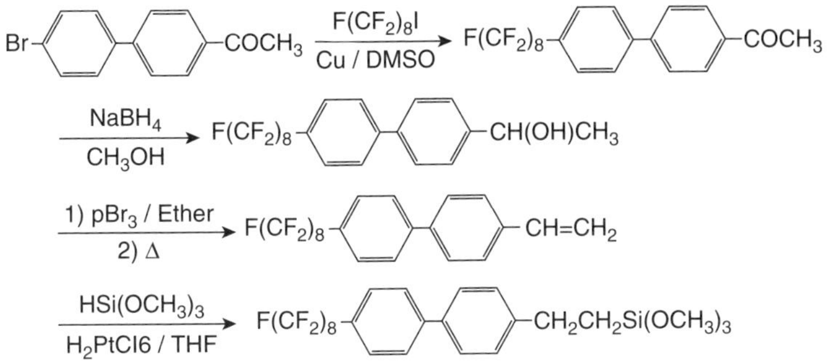

Br–(C6H4)2–COCH3 → [F(CF2)8I, Cu / DMSO] → F(CF2)8–(C6H4)2–COCH3

→ [NaBH4, CH3OH] → F(CF2)8–(C6H4)2–CH(OH)CH3

→ [1) pBr3 / Ether, 2) Δ] → F(CF2)8–(C6H4)2–CH=CH2

→ [HSi(OCH3)3, H2PtCl6 / THF] → F(CF2)8–(C6H4)2–CH2CH2Si(OCH3)3

Scheme 5.1

ring structure into the conventional fluorinated silane coupling agent molecules to improve the thermostability of the modified surface [53–55].

5.3.6.3 Synthesis of Thermostable Molding Agents

Scheme 5.1 shows the synthetic route of a fluorinated silane coupling agent with a biphenyl structure, $F(CF_2)_8(C_6H_4)_2C_2H_4Si(OCH_3)_3$.

The initial water contact angle on the substrate surface modified with this compound was 110°. Even after being exposed to air at 350 °C for 2 h, the modified glass retained a water contact angle of over 100°, showing that the glass still keeps a thermostable low free energy surface. However, the synthesized silane coupling agent was obtained as a mixture of two structures, α- and β-adducts as shown in Scheme 5.2, and the mixture was hardly separated by distillation into the two components. Moreover, the synthesis performed at high temperatures to raise the yield produced a high mixing ratio of α-adduct, while that conducted at low temperatures gave a high mixing ratio of β-adduct at a very low yield. The thermostability of glass surface modified with a mixture obtained at a high temperature differed inappreciably from that modified with the conventional molding agent, $F(CF_2)_8CH_2CH_2Si(OCH_3)_3$.

5.3.6.4 Fluorinated Silane Coupling Agents Thermostable up to 400 °C

Recently, we have succeeded in synthesizing novel compounds with a structural formula of $Rf(C_6H_4)_2CH_2CH_2CH_2Si(OCH_3)_3$ (nF2P3S3M). While details of the synthetic method of coupling agents are not given here because of limited space, their thermostability varies with the length of fluorocarbon

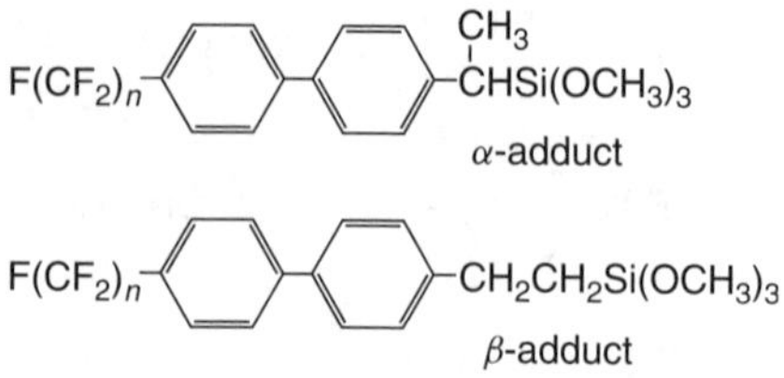

$F(CF_2)_n$–C_6H_4–C_6H_4–$CH(CH_3)Si(OCH_3)_3$

α-adduct

$F(CF_2)_n$–C_6H_4–C_6H_4–$CH_2CH_2Si(OCH_3)_3$

β-adduct

Scheme 5.2

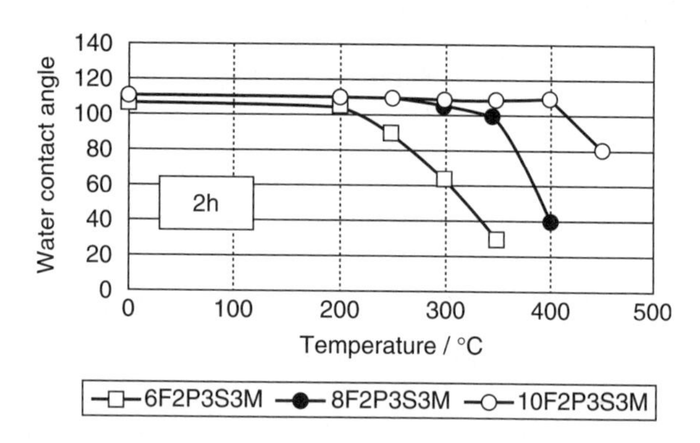

Figure 5.49 Thermostability of glass surface modified with *n*F2P3S3M

chain (Rf) and the compound with Rf = $F(CF_2)_{10}$ is excellently thermostable up to 400 °C. A comparison is made in Figure 5.49 among the thermostabilities of glass surfaces modified with silane coupling agents having C_6F_{13}-, C_8F_{17}-, and $C_{10}F_{21}$- after being exposed to various temperatures for 2 h.

Figure 5.50 shows the results of a comparison between the thermostability of glass surface modified with $F(CF_2)_8CH_2CH_2Si(OCH_3)_3$ used as molding agent and that modified with the newly synthesized $F(CF_2)_{10}(C_6H_4)_2CH_2CH_2CH_2Si(OCH_3)_3$. The surface modified with the latter agent was highly thermostable even after being exposed to air at 400 °C for 10 h (showing no sign of water contact angle lowering), while no thermostability was observed at 350 °C and the water contact angle decreased abruptly for the surface modified with the former agent.

The novel silane coupling agent forms a siloxane network in the vicinity of a modified substrate surface in which the coupling agent molecules bind each other, as do the conventional silane coupling agents on the modified surface (Figure 5.51(a)). Molecules of the coupling agent having two aromatic rings form a modifying layer where the molecules bind each

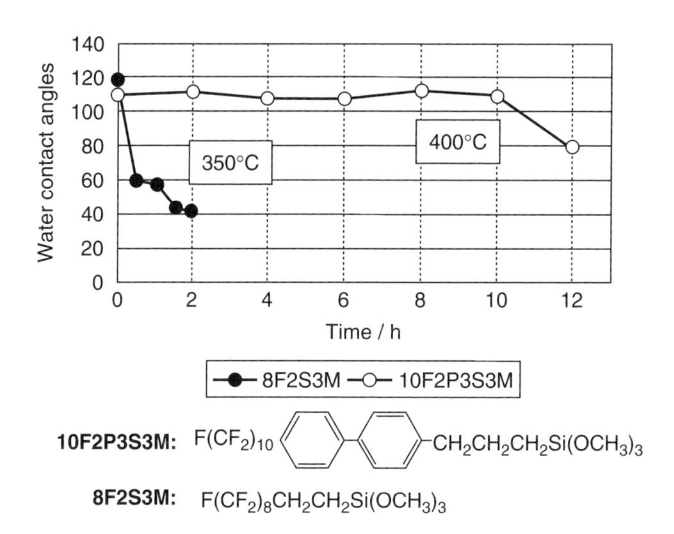

Figure 5.50 Comparison between thermostability of surface modified with $F(CF_2)_{10}(C_6H_4)_2CH_2CH_2CH_2Si(OCH_3)_3$ and that modified with $F(CF_2)_8CH_2CH_2Si(OCH_3)_3$

other closely and solidly through the interaction between their aromatic rings (Figure 5.51(b)). In addition, even when the substrate surface has fine unevennesses, the two aromatic rings of the molecule can arrange themselves along the unevenness and the intervening interaction continues to act between the neighboring molecules to attract each other. Moreover, fluorocarbon chains at the terminals of the coupling agent molecules on the outermost layer of the substrate surface densely cover the surface through hydrophobic interaction (Figure 5.51(c)). These factors contribute to the improved thermostability and moldability of the new coupling agent. The thickness of the coupling agent layer is estimated to be about 2.5 nm on the basis of the molecular chain length.

The mold modified with $F(CF_2)_{10}(C_6H_4)_2CH_2CH_2CH_2Si(OCH_3)_3$ was highly moldable and allowed to transcript from a mold with an aspect ratio of about 10. Figure 5.52 shows an example of nanoimprinting using this mold.

5.3.6.5 Summary

We have described what silane coupling agents are and explained the mechanism by which the agents modify the surface of the substrate. Also, we

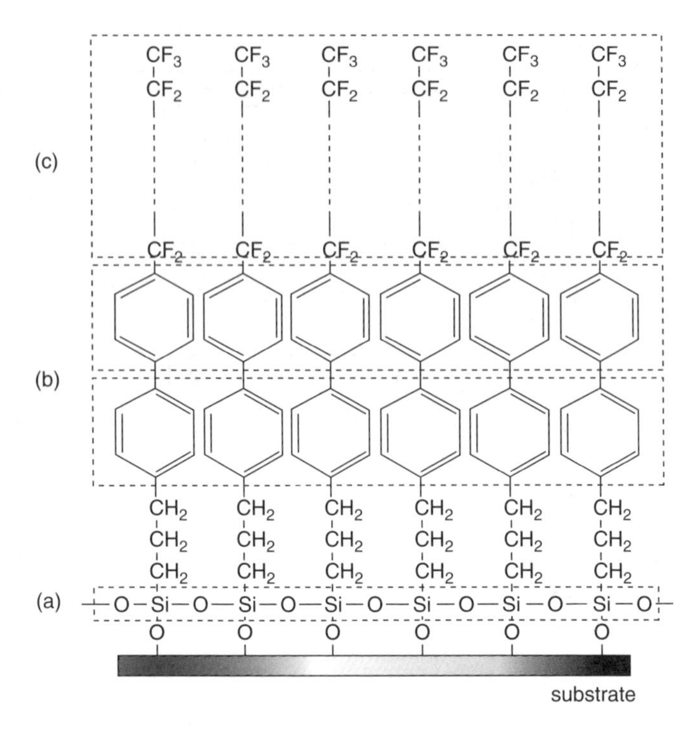

Figure 5.51 Schematic representation of binding a new silane coupling agent to the substrate surface

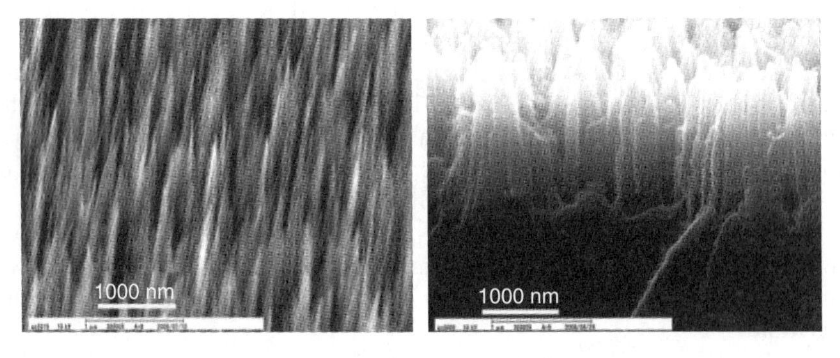

Figure 5.52 Mold surface with moth-eye structure (left) and imprinted polymer surface (right): aspect ratio = 10

have dealt with new molding agents being investigated. Thus, novel silane coupling agents with a biphenyl group introduced in their molecules were synthesized and the thermostability of the glass surface modified with the newly synthesized coupling agent was evaluated. A significant improvement in thermostability was achieved and a high moldability was found to emerge. In particular, the molding agent $F(CF_2)_{10}(C_6H_4)_2CH_2CH_2CH_2Si(OCH_3)_3$ showed a surprisingly high thermostability as an organic surface-modifying agent. In fact, the water contact angle remained unchanged on the glass surface modified with the agent, even after the surface was exposed to air at 400 °C for 10 h. It is believed that this thermostable silane coupling agent can contribute to the improvement in molding duration of UV nanoimprinting and is expected to be widely applicable because the agent is the only molding agent for nanoimprinting to resins with high softening temperature (thermo-nanoimprinting) at present.

5.4 Evaluation Method

5.4.1 Macro Evaluation Technique of Nanoscale Pattern Shape and Evaluation Device

5.4.1.1 Starting Point of Macro Observation Technique

In the method of observation of the pattern shape and evaluation, there are micro and macro techniques. The former uses a microscope. The pattern shape of a microscale area can be observed by optical microscopy. The pattern shape of a nanoscale area can clearly be observed by SEM and AFM. The micro technique is effective in observing the pattern shape of an extremely narrow area, but when the observation view is 1 mm^2, 10,000 operations must be repeated to observe the entire 100 cm^2 substrate. Therefore, it is in fact difficult to observe the entire area by the micro technique due to restrictions of time and cost. In contrast, it is easy to evaluate the uniformity of the pattern shape by observing the entire substrate pattern with a macro technique. This technique does not visualize each pattern shape, but observes the difference in quantity of reflected light in each pixel, and detects non-uniformity of the entire pattern.

5.4.1.2 Development of Macro Observation Technique (from visual observation to automatic measurement)

The inspection technology that uses the macro technique has developed to inspect the uniformity of the resist pattern shape in the photolithography process of the semiconductor. The reason is that the defect of the resist pattern shape is one of the causes of a fatal problem in the manufacturing process of the semiconductor. Therefore, a technology that was able to inspect the uniformity of the entire pattern on the surface of the wafer in a short time (one minute or less) was necessary. Skilled inspection personnel judged the uniformity of the resist pattern on the wafer by using visual inspection equipment until the circuit line width generated became about 0.3 μm (see Figure 5.53(1)). This is the technique used to observe various reflected light from the surface of the wafer. It is judged that there is a non-uniform patterned area if non-uniform reflected light exists. The general technique of observing the diffracted light is described as follows. If white light is illuminated on the surface of the wafer with line and space (L&S) resist pattern, the diffracted light that depends on the L&S pitch and shape is generated from the surface of the wafer. And, if these are the same, the diffracted light of the same color can be observed even in a different area. But, if an area of different pattern pitch or shape exists, then diffracted light of a different color is observed (see Figure 5.53(2)). Thus, if the difference in diffracted light is considered, a defective area of the pattern shape could easily be detected. However, visual inspection became difficult when the semiconductor technology changed to nanoscale generation. Then, the automation technology of macro inspection came to be developed. The feature of this technology is to capture images

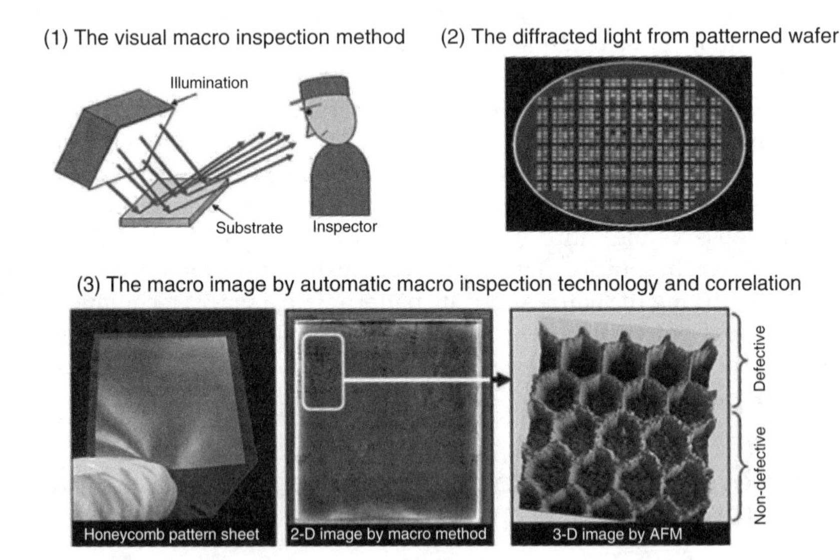

Figure 5.53 Visual macro inspection and automatic macro inspection

using an electronic camera equipped with high-sensitivity photosensor. It is also possible to detect slight reflected light that is not visible, and to use UV rays. As a result, it became possible to detect infinitesimal non-uniformity of the entire pattern. The advantage of the automation is that quantification is possible. Quantification by the macro technique defines the uniformity of the entire pattern (for example, differences in line width, pitch, height, and pattern shoulder shape, etc.) by the value of the contrast and color, and turns it into numeric data. Therefore, macro image data can correlate with micro image data (for example, SEM and AFM images). When an area of the macro image of a non-uniform honeycomb pattern from the nanoimprint process is observed with AFM, the defective shape of the honeycomb can be seen. It is supposed that there is a correlation between the change in brightness of the macro data and the change in pattern shape (see Figure 5.53(3)). The correlation of the macro method and the micro method is described in detail in the next section.

5.4.1.3 Measurement of Pattern Shape by the Macro Method

In this section, two examples describe differences in pattern shape on the nanoscale, measured to high sensitivity. The first example is the detection of a minute change in line width. A resist pattern of design value hp 80 nm is formed on the surface of an 8-inch wafer as shown in Figure 5.54(1). There is a difference in line width of about 5 nm in each area, with 4 mm pitch. The optimized macro illumination is irradiated on the surface of the wafer with resist pattern, and the image of the reflected light is captured using a CCD camera at a specific angle. Between the brightness profiles of the macro image and the line width value of the resist pattern (measured by SEM), there is an extremely high correlation $R^2 = 0.995$ (see Figure 5.54(3)). The second example is the minute difference in shape measured for the edge shoulder of the pillar resist pattern. A design value 300 nm cube is formed, with 600 nm pitch on each die of the surface of a 3-inch wafer. The radius of curvature of the shoulder on each die after development is different, since four die are exposed to the pattern in different doses by the electron beam lithography device (see Figure 5.54(2)). The optimized macro illumination is irradiated from a low angle to the surface of the wafer, and the image of the reflected light from the resist pattern shoulder is captured by CCD camera at a specific angle. Between the brightness profiles of the macro image and the curvature radii on the edge shoulder of the resist pattern (measured by SEM), there is an extremely high correlation $R^2 = 0.975$ (see Figure 5.54(4)). A relative change in pattern shape can be measured as non-uniformity of the brightness of the macro image without measuring the individual pattern shape when the latest macro imaging technologies are used.

(1) Detection of difference of the pattern width by macro imaging technology.

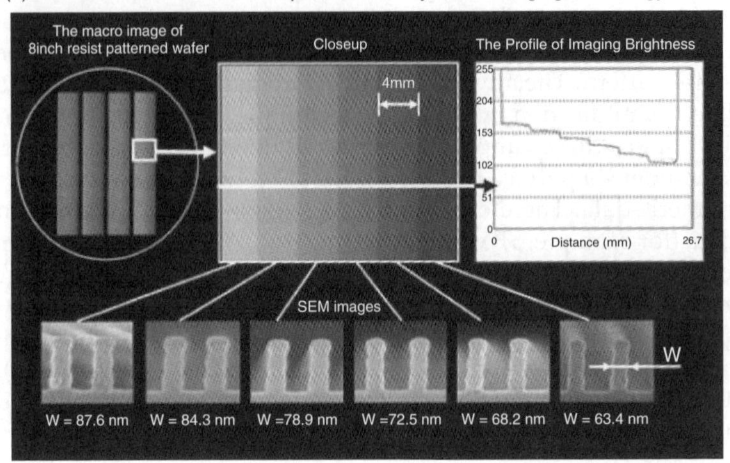

(2) Detection of difference of the pattern shape by macro imaging technology.

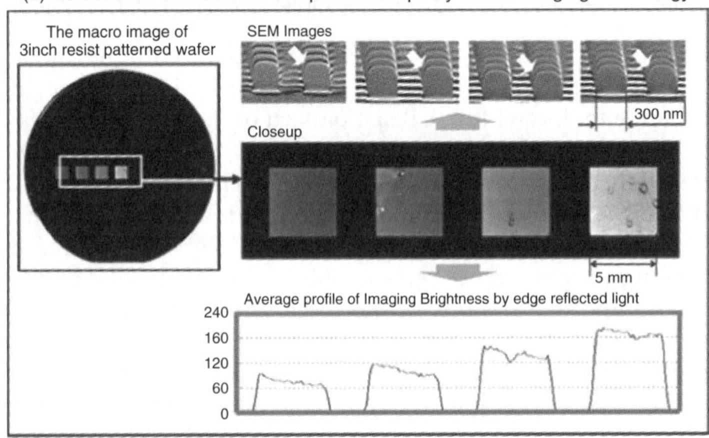

(3) Correlation of resist pattern width

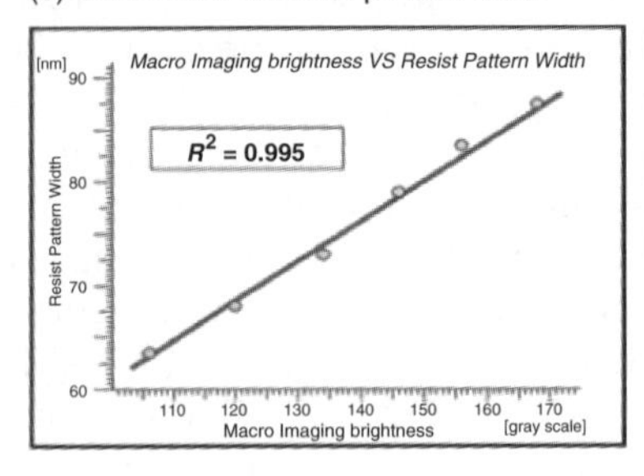

(4) Correlation of resist pattern edge shape

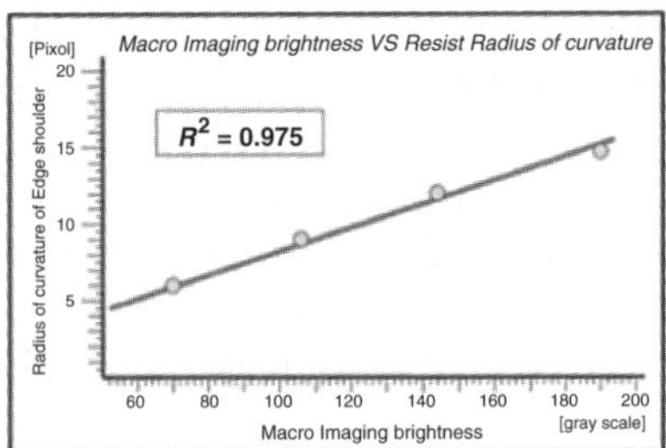

Figure 5.54 Sensitivity and correlation by the macro method

5.4.1.4 Evaluation of Patterning Uniformity and Fidelity, and NIL Device

The macro method can be used to evaluate the following NIL processes.

(1) Uniformity, fidelity, and accuracy of master mold.
(2) Replicating from the master mold to the working mold.
(3) Patterning the resin film by the mold.
(4) Comparison of plural resin films (same mold used).

In this section, we outline the macro evaluation device that enables the above-mentioned evaluations and one sample of the macro evaluation method. The device has the following features (see Figure 5.55(1)).

(1) The scanning mechanism of the substrate and the image can be taken with a linear CCD sensor camera synchronized with the scanning speed.
(2) The linear CCD sensor camera can be positioned on a circular arc orbit, and the image of the reflected light can be captured at various angles.
(3) The macro illumination unit can be set up flexibly at the optimized position.
(4) The image of the same position on the substrate can be captured at any time for any camera angle.

Since this device has a structure such that the specific luminous flux can be selectively detected, the reflected light that shows the feature of the pattern shape can be selected (see Figure 5.55(3)). And, since an actual evaluation device has a structure such that the linear CCD sensor camera installed in the vertical revolute robot can be moved to any position, the optimum optics condition can be set (see Figure 5.55(2)).

Next, two resin films with the same honeycomb pattern (sample L and sample R) are evaluated by self and mutual pattern uniformity using this device.

1st process: The linear CCD sensor camera is positioned from 30° to 90° in 1° pitches, and the device captures 61 macro images (see Figure 5.55(4)).

2nd process: Each image is divided into 20 areas, and the non-uniformity value in each area is calculated. In this example, the standard deviation is used as the value of non-uniformity. Comparing the correlation of L3 and R3, and the correlations of L20–R20, the correlation coefficient of the former is 0.87 and of the latter is 0.10 (see Figure 5.55(5)).

3rd process: All the correlation coefficients between sample L and sample R are graphed. This is an evaluation result for the comparison of pattern

(1) The concept of the evaluation device (2) The actual device

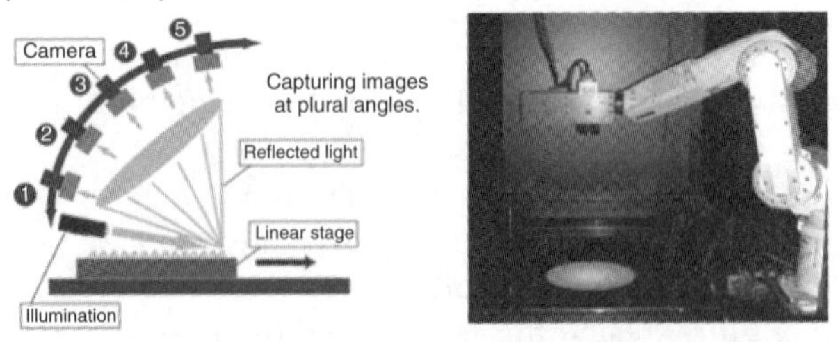

(3) Edge reflected light detection at each camera angle

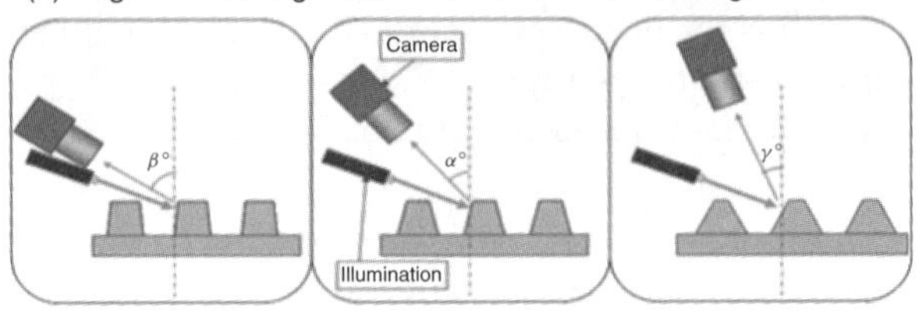

(4) Multimode non-uniformity detection method

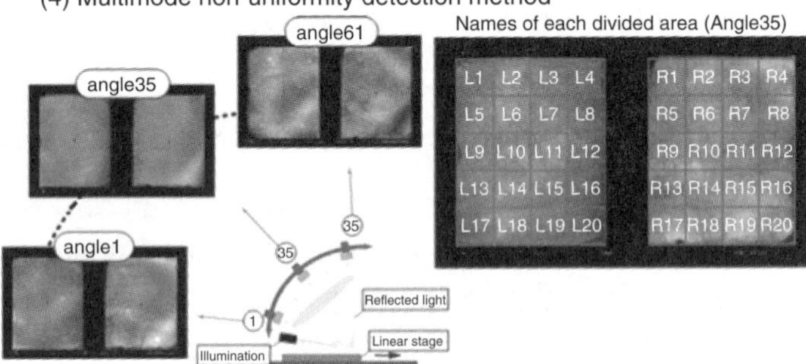

(5) Comparison of reflection intensity between Area-L and Area-R

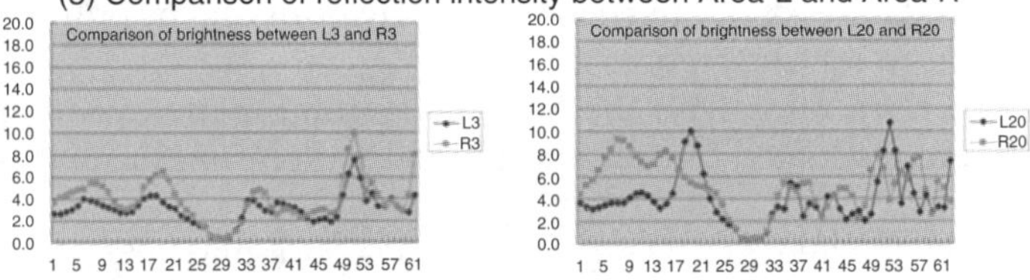

(6) Correlation coefficient between Area-L and Area-R

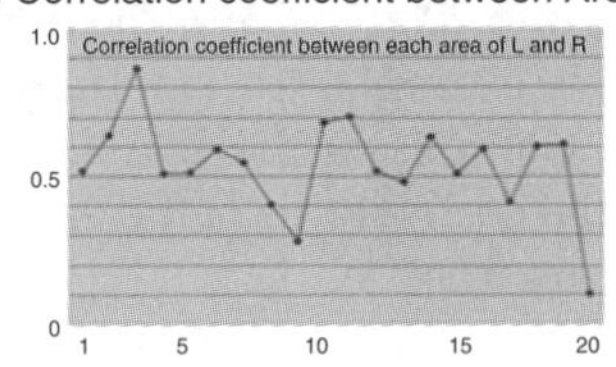

Figure 5.55 The macro evaluation device and the correlation of nanoimprint sheets

transfer for two samples, and the following points can be recognized from this graph (see Figure 5.55(6)).

(1) The correlation of outer areas is low (L8, L9, L13, L17, L19).
(2) The correlation of center and upper areas is high (L2, L3, L10, L11, L14).

The outer part of the substrate can be judged to have a large difference in pattern transfer, although the center part is equal in the comparison of the two samples. A macro image with high uniformity is captured if the pattern transfer is excellent, since this substrate is a repetition structure of the honeycomb pattern. Therefore, if this evaluation is repeated while changing the conditions of the imprint process, the pattern transfer condition with the fewest differences might be best. However, it should be evaluated to guarantee the uniformity of the mold. Moreover, several evaluations are fundamentally possible, as follows.

(1) Prediction of the lifetime of the mold (periodic monitoring of macro image data).
(2) Contamination and scratching on the surface of the mold.
(3) Deterioration of the mold release.
(4) Common defects on the NIL sheets.
(5) Trouble with NIL manufacturing equipment and the manufacturing process.
(6) Process control and yield management.

In this section, we have shown how to evaluate a nanoscale pattern by the index of uniformity when the macro method is used. This method is an extremely effective evaluation technology of pattern uniformity over the entire area, although it is not fundamentally effective in a narrow area. The development of various evaluation technologies for pattern uniformity can be expected in the future using macro-imaging technology that captures macro images of high quality.

5.4.2 Characterization of Photocurable Resin for UV Nanoimprint

The requirements for photocurable resin are classified into two categories: basic process characteristics that are independent of applications, and application-oriented characteristics (Figure 5.56). When materials are developed, it is important to improve their characteristics, of course, but we must also develop the techniques of characterization.

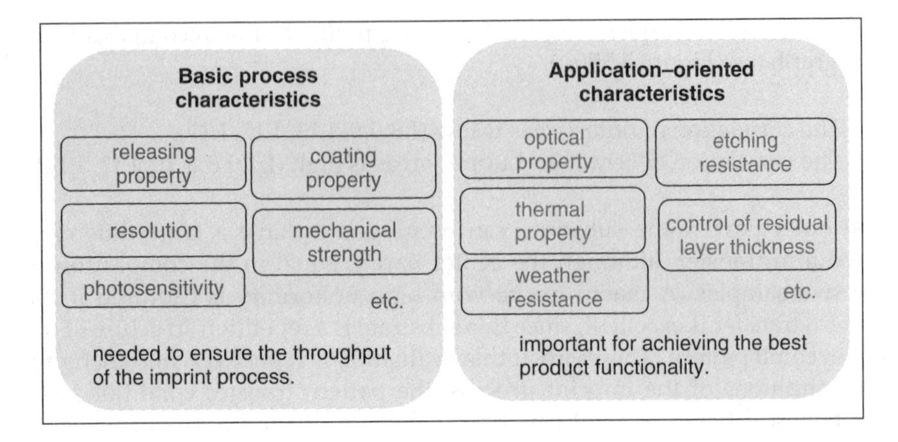

Figure 5.56 Demand characteristics for UV-NIL resin

5.4.2.1 Basic Process Characterization

Techniques for basic resin process characterization will be described. The basic process characteristics of UV-NIL resin are needed to ensure the throughput of the imprint process. They include coating properties, viscosity, releasing properties, mechanical strength, resolution, reproducibility, and photosensitivity (reaction rate, extent of polymerization). Here, we discuss the releasing properties, transfer reproducibility, mechanical strength, and conversion rate.

(1) Releasing properties
In the releasing process of UV-NIL, there is a type of failure in which the resin pattern is damaged [56] because resin tends to adhere to and may not be separable from the mold. For this reason, a release-promoting treatment is usually applied to the mold in advance. In the case of a quartz mold, for example, the mold surface is usually water-repellent finished using a silane coupling agent of the perfluoroalkyl group [57, 58]. The greatest practical problem in the releasing process is the level of release-enhancing capability and its stability [59, 60].

First, reports are described on the effect of releasing treatment [59]. A sample and jig were fabricated for the measurements, as shown in Figures 5.57 and 5.58. The stress needed to separate two slide glass plates was measured, and the effect of releasing agent was studied. The adhesive force is measured using PAK-01 (Toyo Gosei [61]). The releasing agent for UV-NIL is Optool

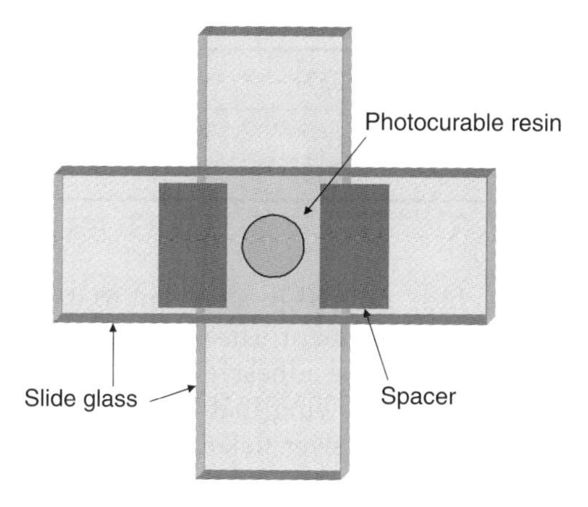

Figure 5.57 Experimental setup of sample for measurement

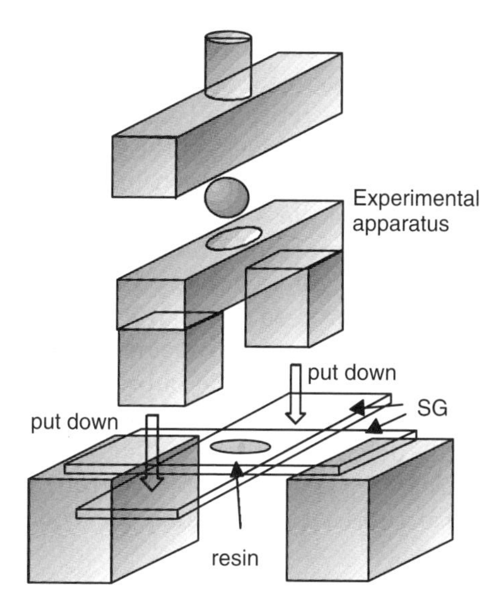

Figure 5.58 Experimental setup of measurement for adhesive force and durability

Table 5.5 Measurement results of adhesive force of PAK-01

Dilution ratioof release agent	0.05%	0.1%	0.2%	Untreated
Mean value (MPa)	0.053	0.053	0.060	0.83
Standard deviation	0.019	0.023	0.020	0.13

DSX (Daikin Industries, Ltd), diluted in a solvent to 0.05–0.2% by weight; dipping is used as the release treatment. The measurement results are shown in Table 5.5. It is confirmed that the adhesive force is greatly reduced by the agent at any level of dilution, showing that the releasing treatment makes release easy. Note that there are fewer flake failures when the difference is greater in the adhesive force between the mold and resin than the adhesive force between the resin and substrate. Compared with the case of untreated glass plates, which can be taken as representing the untreated case of adhesion between substrate and resin, the adhesive force of the treated mold is about 10 times greater. This gives a sufficiently great contrast.

Second, reports on the stability of the releasing treatment will be described. Many reports show that the effectiveness of the releasing treatment gradually degrades as the nanoimprint is repeated [62, 63]. For example, Houle *et al.* find that the degradation in effectiveness of the releasing treatment depends on the properties of the photocurable resin [62]. However, note that the degradation still occurs, regardless of whether the resin is superior or inferior. It is proposed that a metal thin film over the silane coupling agent can be a promising releasing agent, as discussed earlier. Tada *et al.* studied the degradation in the releasing process using an X-ray analysis technique [63]. They found that the mixture ratio of the elements changes in the releasing agent and the thickness decreases as the imprinting is repeated, and they found experimentally that the radical supplied from the photopolymerization agent does not directly deteriorate the releasing agent. The mechanism of degradation of the releasing capability is not yet fully known, and remains an important technical subject in promoting the wide commercial use of nanoimprint technology.

In this experiment, the imprinting is repeated using a release-treated quartz mold, and the relation is examined between the adhesive force during the release process and the contact angle of water on the mold surface. A quartz plate is used as a mold with a transfer surface of 25 mm × 25 mm without patterns, and two types of sample – with and without release treatment – are prepared. In the releasing process, perfluorooctylethyltrichlorosiline ($C_8F_{17}C_2H_4SiCl_3$, Glest) is used as a releasing agent in the vapor phase treatment (Figure 5.59). The contact angle of water on

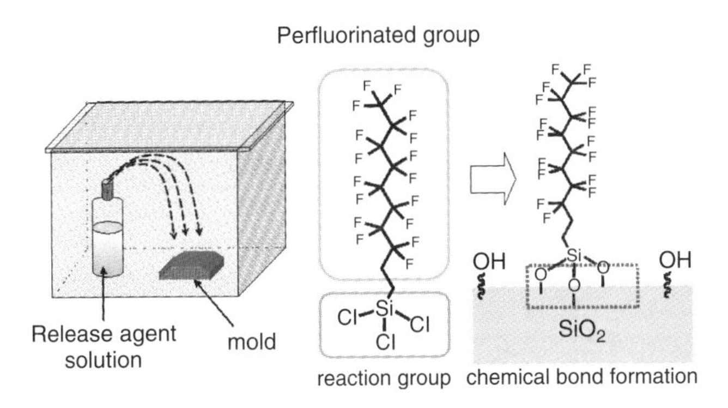

Figure 5.59 Treatment method for mold surface

the mold surface is found to be 40° without treatment and 80° with release treatment. Using these types of mold, the imprinting is repeated using an imprint apparatus from Mitsui Denki Seiki. The process of the repeated imprinting will be explained next. Firstly, resin is spin-coated with a finished thickness of 200 nm on the silicon wafer substrate. Then, the mold is contacted with the resin film to apply pressure using the imprint apparatus. After the set weight of 200 N is reached, resin is cured by UV irradiation ($100 \, mJ/cm^2$), during which the weight is kept constant. The mold is released from the resin after UV irradiation. Then, the resin residue on the mold is checked, along with the adhesive force and the contact angle of water on the mold surface. The resin residue on the mold is checked visually, and the contact angle of water is measured using a static contact angle meter. The adhesive force shows a negative value in the earlier stage of the releasing process, and the value reduces to zero when the releasing process ends. The maximum of the negative values was treated as the adhesive force. The measurement result is summarized as follows. Some of the resin begins to be left on the untreated mold after imprinting has occurred several times, while resin residue is not observed in the release-treated mold. We now discuss the variation of adhesive force and contact angle. Using photocurable resin and a mold with contact angle of 80°, Figure 5.60 shows the variation in adhesive force and contact angle up to the 1000th transfer. The adhesive force of the resin remains invariant up to 1000 transfers, and the resin does not stick anomalously to the mold. However, the contact angle on the mold decreases very slowly as the transfer is repeated, indicating that the variation in adhesive force does not equate with the variation in contact angle.

The decrease in contact angle probably occurs because a wider surface area of quartz is exposed owing to the decreasing amount of releasing agent

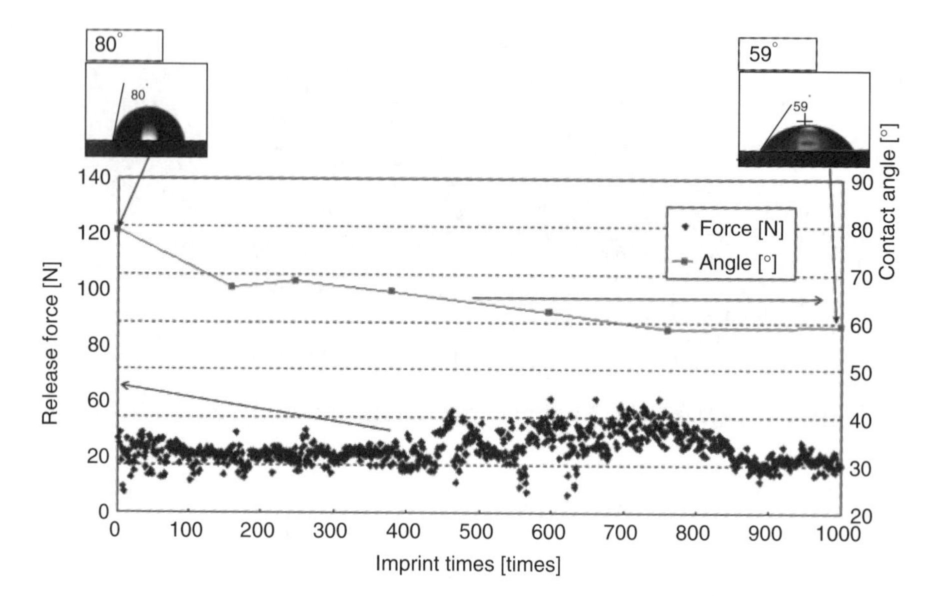

Figure 5.60 Experimental results on durability of release layer

on the mold surface. It appears necessary to discuss the releasing properties not only from the viewpoint of the contact angle and adhesive force, but also from several other viewpoints.

(2) Adhesion to the substrate
In another type of UV nanoimprint failure, resin comes off the substrate or underlayer abnormally. This problem is handled not only by improving the release capability, but also by improving the adhesion properties between the resin and substrate. In relation to this, there is a report on the debris appearing in a UV nanoimprint, where six types of commercial resin and three types of substrate are cross-examined to see if debris occurs [64]. Note that the affinity between the resin and the substrate is closely related to failure with the appearance of debris.

(3) Efficiency of conversion
Resin for UV-NIL is cured by UV irradiation. To improve productivity, it is desirable for the resin to be cured by the smallest amount of UV irradiation. The polymerization resin can be evaluated by monitoring the cure process either through the change in mechanical properties [65] or through chemical analyses [65–67]. The reaction rate is measured using

Fourier transform infrared spectroscopy (FT-IR) in the chemical analyses of the polymerizable functional groups in the resin. The measurement method is as follows. In the FT-IR measurement, the resin sample on the silicon wafer is UV-irradiated with optical intensity 5 mW/cm^2. The change in absorption intensity of polymerizable groups in resin (with peaks at 810 or 1630 cm^{-1}) is tracked in real time, and the conversion rate is calculated from the change (Figure 5.61). Equation (5.1) is used to calculate the extent of conversion at time t. By taking an arbitrary baseline at time t at the peak position of the polymerizable group, we define 0% conversion by the peak or area before irradiation, and we define 100% conversion by the peak or area after, as shown in Figure 5.61.

$$\% \text{ conversion} = (A_0 - A_t)/A_0 \times 100 \tag{5.1}$$

where A_0 is the peak or area before UV irradiation, and A_t is the peak or area after irradiation time t.

Figure 5.62 shows the measurement result for the conversion of PAK-01. PAK-01 has a large initial reaction velocity and the saturated extent of conversion is as large as ≈85% for exposure of 5 mJ/cm^2, and 90% for 10 mJ/cm^2. Since the polymerization of resin is greatly affected by the choice of monomer, oligomer, and photopolymerization initiator, a measurement result like this can be used in the selection of materials.

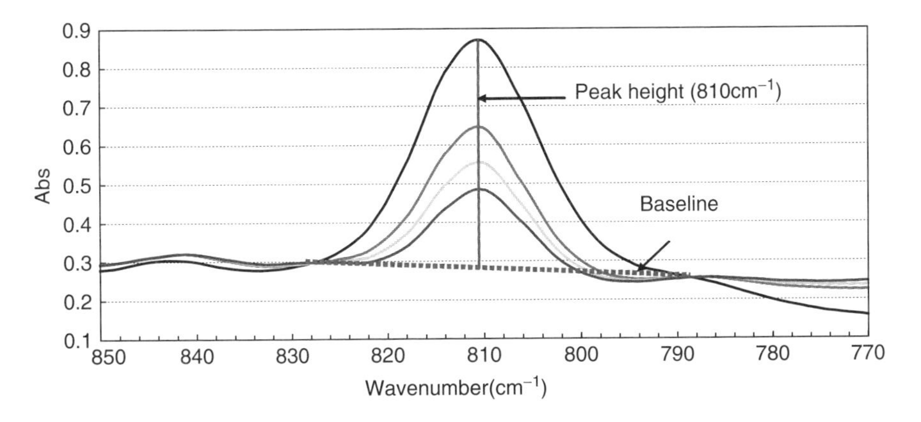

Figure 5.61 Method for estimation of reaction value

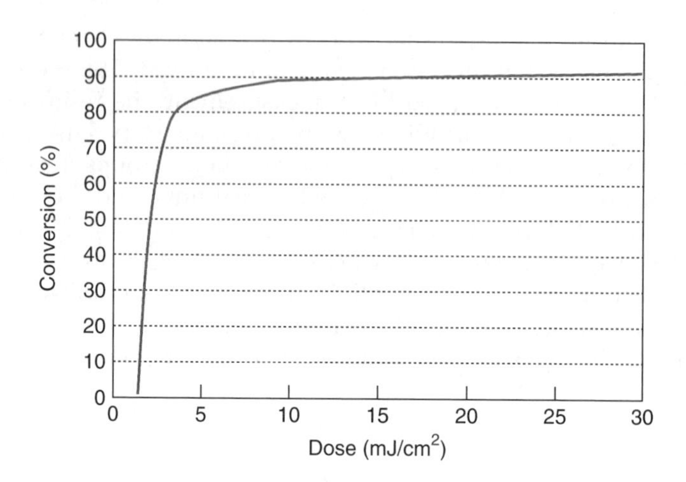

Figure 5.62 Measurement result of photo-curing rate of PAK-01.

(4) Transfer accuracy
The transfer accuracy of UV-NIL is evaluated by comparing the pattern configurations in the mold and in the resin. While the resolution of optical lithography depends on the wavelength of the light source, the resolution of UV-NIL depends on the mold configuration.

There are various tests to evaluate resolution; two of them are taken up here. First, studies on the minimum resolvable pattern size will be described [68, 69]. Hua *et al.* attempted to transfer a carbon nanotube configuration. First, the carbon nanotube configuration was transferred to the polydimethylsiloxane of thermal cure type. Then, it was imprinted onto photocurable resin using the cured polydimethylsiloxane as mold. As a result, a carbon nanotube configuration of 2.4 nm was successfully transferred to the photocurable resin.

We now discuss the technique of transfer accuracy evaluation where the difference in configuration is compared between the mold and the resin [70, 71]. Hiroshima reported on the comparison of line edge roughness (LER) in the patterns of the mold and transferred resin [71]. The SEM measurement gave an LER difference of 0.1–0.2 nm between the mold and the transferred patterns. This value is excellent compared with those of other processing techniques, indicating that the resolution of the nanoimprint is surprisingly good. While only a small number of reports are available on the relation between the resolution and the resin type, there are some reports on the relation between the resin type and the cure shrinkage [72].

(5) Viscosity
Nanoimprinting is a molding process technology, and the viscosity of resin influences the processing time. Lowering the viscosity reduces the processing time, which is advantageous in industrial applications. The viscosity is $\approx$1 mPa s to apply vinyl ether monomers and 3–4 mPa s for acrylate monomer [67].

5.4.2.2 Application-Oriented Characterization

The characteristics of UV-NIL resin are important for achieving the best product functionality. While many good characteristics are required, note for example that the imprint process can be classified into two categories: the removal of resin residue in the concave portion of the imprint pattern may or may not be required, depending on the application. Applications where this is not required, which are energetically studied by making use of the nanostructure of the resin surface, include optical devices. Imprinting becomes easier if the residual layer thickness is greater, and a thickness on the order of micrometers is usually adopted. Since the cured resin remains in the device, its characteristics will greatly influence the device reliability. Therefore, the cured resin must have an excellent level of weather resistance. In contrast, the former applications include those for lithography using such wet processing as dry etching, plating, and lift-off. In applications for lithography, complicated processes usually follow the imprinting, so the key to success is consistency with these subsequent processes. In particular, the residual thickness greatly affects the subsequent processes, and it is important that the resin film be uniform and as thin as possible [73].

(1) Characterization of resin for permanent devices
Issues relating to resin depend on the target device, and will include: resistance to climatic conditions, resistance to humidity, thermal properties, and optical properties (transparency, refractive index, etc.). There is a variety of device types, and it is supposed that unique approaches are individually pursued.

Studies on thermal properties and transparency will be described in the following. The nanopattern fabrication on flexible resin films will also be described.

(2) Thermal properties
The thermal properties of resin are among the items that are very important when resin is used in permanent devices [59, 74, 75]. We evaluated

the heat tolerance of cured resin using differential thermogravimetry [59]. The upper temperature limit is defined as the temperature that reduces the weight by 5%. The samples are acrylate resin, epoxy resin, and organic–inorganic hybrid resin. The measurement apparatus is TG-DTA200S (Mac Science), and the temperature is increased by 10 °C/min in nitrogen. The measurement shows that the upper temperature limit is ≈270 °C for acrylate resin, and is also ≈270 °C for alicyclic epoxy resin. It is postulated that the similar level of heat tolerance is probably observed because the alicyclic epoxy resin tested has ester binding in the molecule, just as acrylate resin does. In contrast, in resin where the main constituent is novolac epoxy resin, which does not have ester binding, there is an improved upper temperature limit of 310 °C and, further, the organic–inorganic hybrid resin gives an upper temperature limit as high as ≈340 °C.

(3) Transparency
Optical applications require that the cured resin be colorless and transparent. The coloring of cured resin is strongly related to the composition of the cured material, and a good result is obtained by carefully choosing monomers and photopolymerization initiators that are not causative of coloring. It should be noted that impurities might cause coloring if they are present in the photopolymerization initiator. Figure 5.63 shows the UV spectrum for the cured films of commercial UV nanoimprint resin (PAK-01, PAK-02, Toyo Gosei).

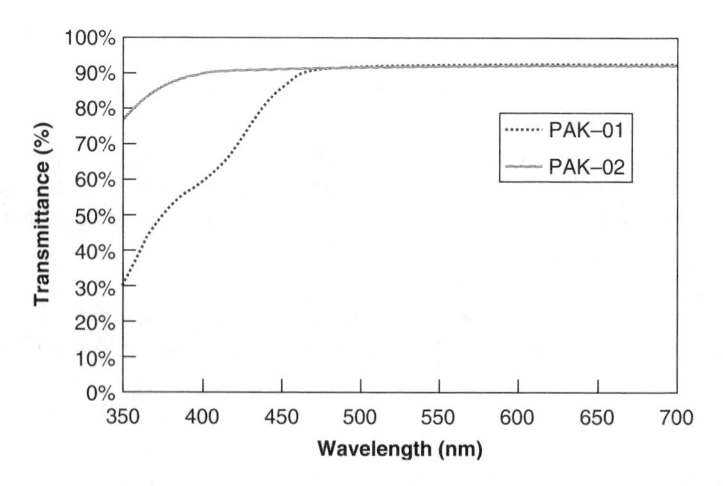

Figure 5.63 Transmission data of cured PAK-01 and PAK-02 (thickness: 50 μm)

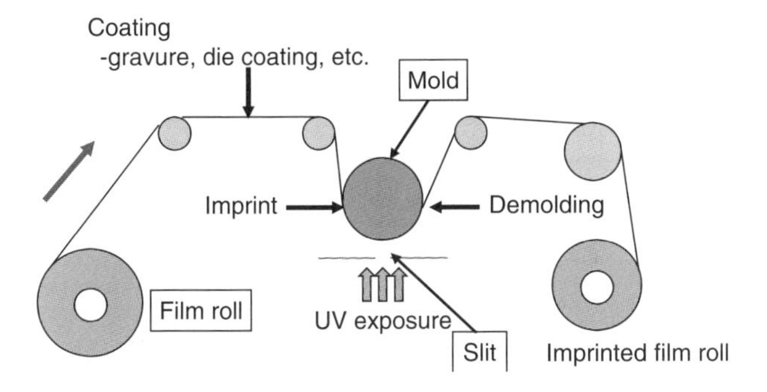

Figure 5.64 Illustration of RTR process

(4) Example of processing
Roll-to-roll imprinting offers high-speed, large-area continuous transfer [76]. This technique is expected to be used extensively in the future, especially for applications that demand low-cost, high-volume production. In particular, applications to functional films for display are gaining attention. An example of continuous transfer using the RTR technique will be discussed in more detail. Here, the roll-type mold is fabricated by attaching electroformed Ni molds on a metal roll. Adhesive-treated PET films (Cosmoshine A4100, Toyobo) with width 21 cm and thickness 100 μm are used as the substrate. The RTR imprinting process is shown in Figure 5.64. The transfer process is as follows. First, the film substrate is coated with PAK-02. Then, the coated resin and mold come into contact with each other, and the mold is filled up with resin to transfer the pattern. UV light is irradiated from the substrate side for photopolymerization, and then the film substrate is released. The above steps are continuously processed with a film feed speed of 6 m/min. Figure 5.65 shows the pattern transferred.

5.4.2.3 Applications for Lithography

(1) Resin for dry etching
Dry etching is widely used in the electronic, including the semiconductor, industry. It contributes significantly to microfabrication, and is very attractive from the viewpoint of nanoimprint technology. From the inception of nanoprinting, using the transferred resin pattern as a mask for dry etching has repeatedly been attempted. The requirement for photocurable resin is tolerability against dry etching. This can be

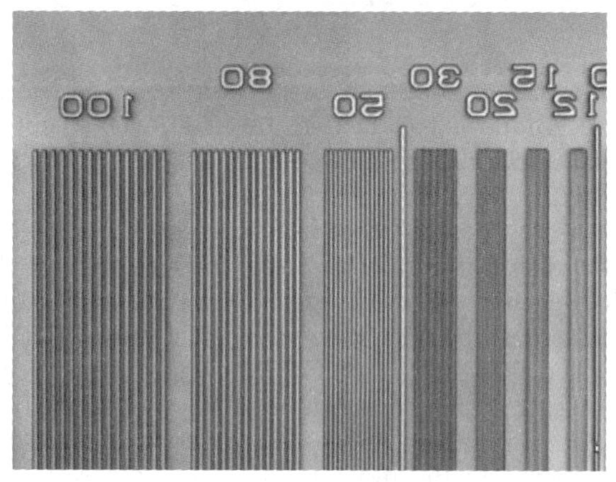

Figure 5.65 Imprinted film sample (upper: appearance of the film, lower: optical microscope image of imprinted pattern)

improved by properly choosing the correct monomers and oligomers for the resin. There is a report on a technique that is based on the improvement of the photoresist technique with respect to tolerability against dry etching [77]. In this report, the Onishi parameter and ring parameter of the constituents of resin are discussed.

An improvement observed by Fukuhara *et al.* [78] will now be described. PAK-01 was tested along with a newly trial-manufactured resin, PAK-TR11. The result of dry etching using these is shown in Figure 5.66. Owing to the difference in constituents, PAK-TR11 has twice the dry-etching tolerability

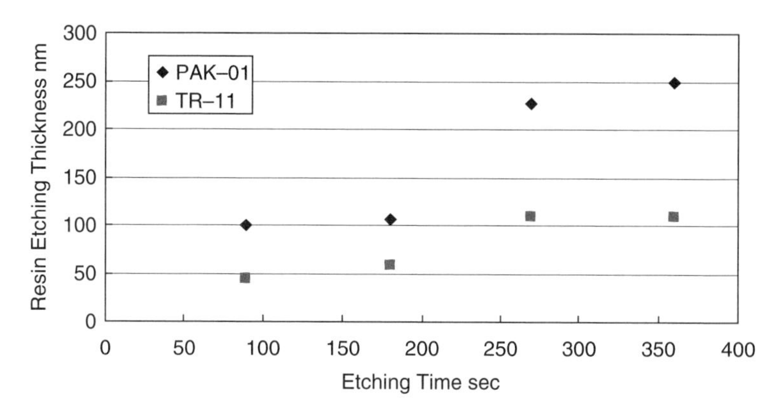

Figure 5.66 Comparison of dry-etching rates by PAK-01 and PAK-TR11 (gas: SF_6, C_4F_8)

of PAK-01. To reduce the residual thickness, dilution with an organic solvent is used, and the thickness of resin after spin-coating and before curing is reduced to ≈60 nm. The silicon wafer is processed using PAK-TR11 (Figure 5.67). By coating the wafer with photocurable resin, a nanopattern of resin is formed using UV-NIL. Then, the residue of the resin pattern is removed using oxygen dry etching, and the silicon wafer is etched with the remaining pattern as a mask. The silicon pattern obtained is shown in Figure 5.68.

The use of a multilayer process is also actively being pursued [79]. The use of a multilayer reduces the effect of irregularity in the substrate, as well as reducing scattering in the residual film thickness. In semiconductor processes, where exact line width control is required [80], the use of a multilayer process will become the mainstream if UV-NIL is adopted.

(2) Resin for wet processes
Plating, etching, and lift-off are the main wet processes in the industry. Techniques to combine them with nanoprinting have not yet been published in abundance. An application for plating will be described [78, 81]. In addition to the residual film thinness, an important item of characterization in the use of resin is tolerability in the plating solution. PAK-TR21 was tested, in which the constituents are adjusted for improvement in this application, and a great enhancement in tolerability was found in the plating solution, such that there were no failures (e.g. swelling and flaking). Figure 5.69 shows the external appearance while immersed in the plating solution.

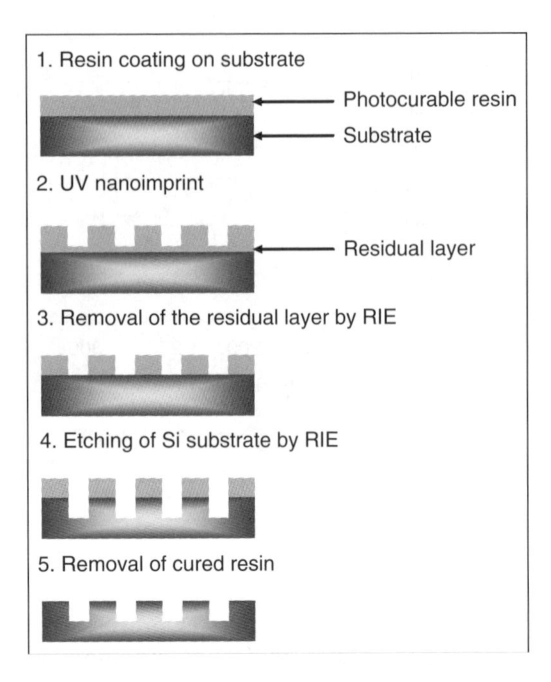

Figure 5.67 Process image of fabrication of Si nanopatterns

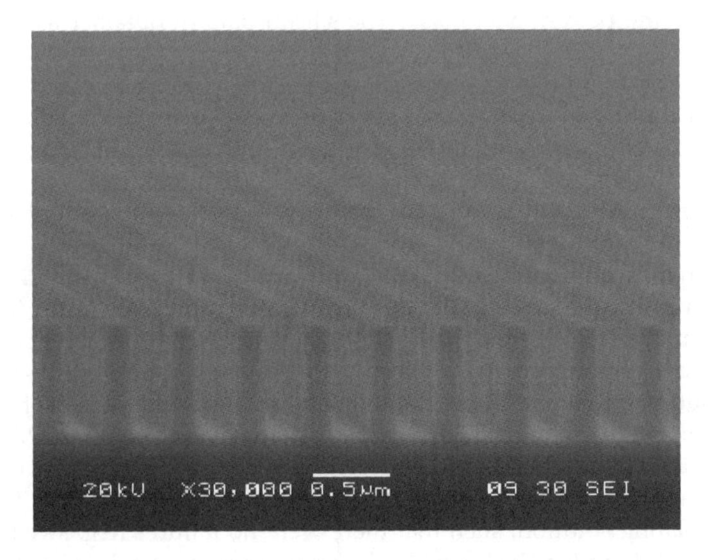

Figure 5.68 SEM image of fabricated Si pattern (140 nm pillar)

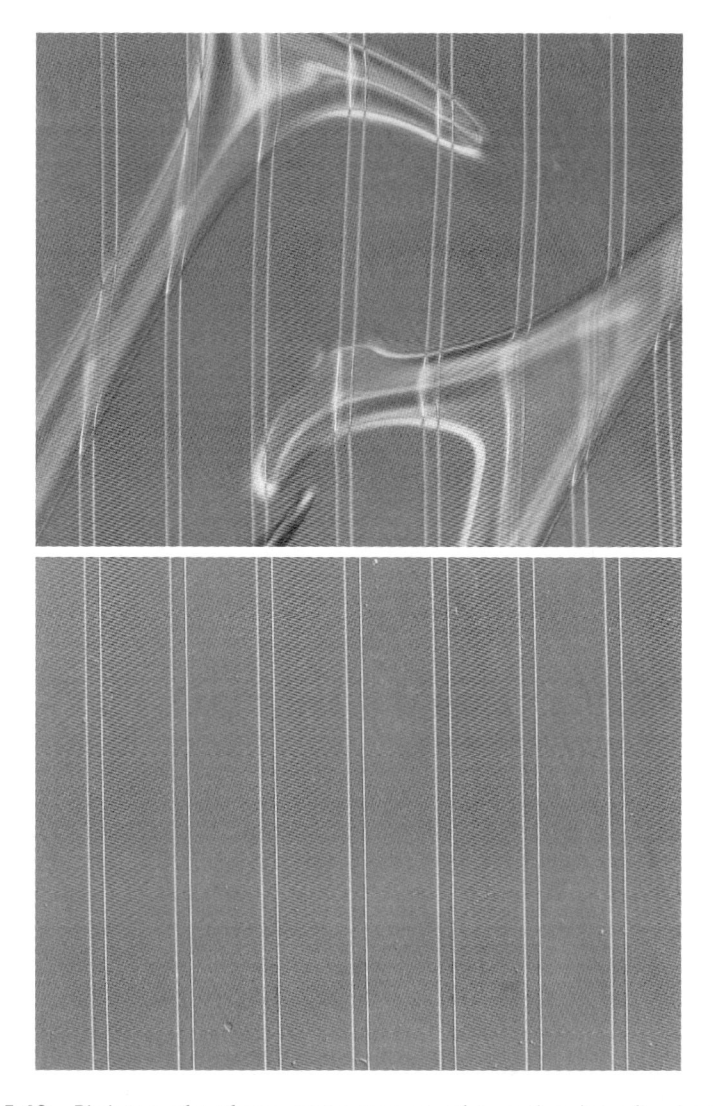

Figure 5.69 Pictures of surface appearance of cured resins after immersion (upper: PAK-01, lower: PAK-TR21)

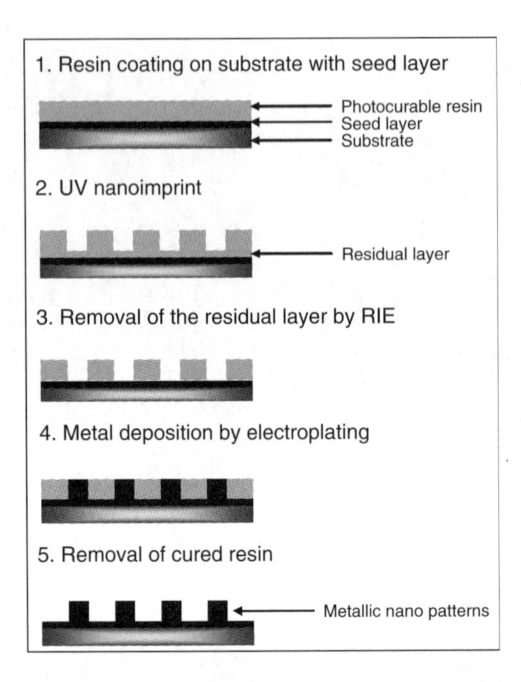

Figure 5.70 Fabrication method of nano metallic patterns

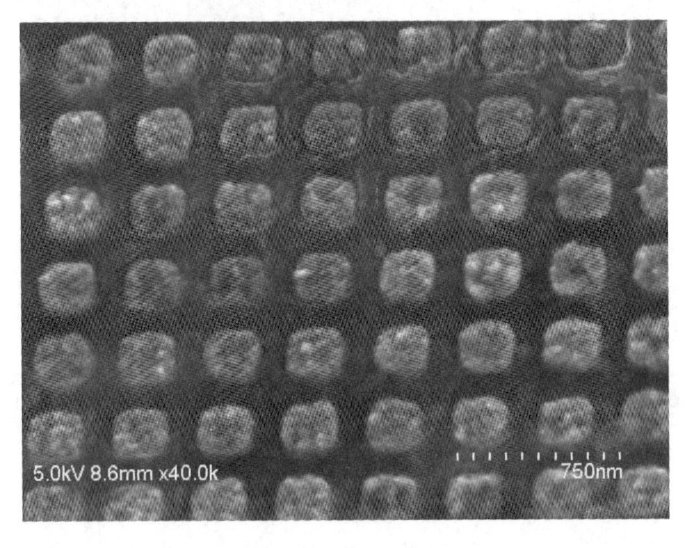

Figure 5.71 SEM image of 240 nm diameter Co alloy patterns fabricated by electrodeposition

Using PAK-TR21, metal nanopatterns are formed using electrolytic plating and UV-NIL. Figure 5.70 shows the process. First, photocurable resin is coated on a silicon wafer, on which a conducting layer has been formed, and the resin nanopattern is formed by the UV-NIL process. Then, the residue of the resin pattern is removed by dry etching, as discussed in the dry-etching section, so that the conducting layer is exposed. Lastly, using electrolytic plating, metal is plated onto the conducting layer to form a metal nanopattern. Figure 5.71 shows the metal pattern formed by the electrolytic plating technique, where the cobalt alloy is plated onto the openings of the resin and then the resin is removed. While other techniques use complicated processes or expensive apparatuses in forming alloy nanopatterns, the adoption of the UV-NIL technique makes the process relatively simple and easy.

References

[1] Kondo, M., Yasuda, H., Kubodera, K., and Fujimori, S. 1976. Japan Society of Applied Physics 37th Autumn Meeting, Vol. **2**, p. 404 (in Japanese).

[2] Kondo, M. and Fujimori, S. 1977. IECE Tech. Rep. CPM76-125 (in Japanese).

[3] Fujimori, S. 2009. Fine pattern fabrication by the molded mask method (nanoimprint lithography) in the 1970s. *Jpn. J. Appl. Phys.* **48**: 1–7 (English review paper of [1, 2]).

[4] Chou, S.Y., Krauss, P.R., and Renstrom, P.J. 1995. Imprint of sub-25 nm vias and trenches in polymers. *Appl. Phys. Lett.* **67**: 3114.

[5] Taniguchi, J., Iida, M., Miyazawa, T., Miyamoto, I., and Shinoda, K. 2004. 3D imprint technology using substrate voltage change. *Appl. Surf. Sci.* **238**: 324–330.

[6] Unno, N., Taniguchi, J., and Ishii, Y. 2007. Sub-100 nm three-dimensional nanoimprint lithography. *J. Vac. Sci. Technol. B* **25**: 2361–2364.

[7] Park, S., Lim, T., Yang, D., Jeong, J., Kim, K., Lee, K., and Kong, H. 2006. Effective fabrication of three-dimensional nano/microstructures in a single step using multilayered stamp. *Appl. Phys. Lett.* **88**: 203105.

[8] Nakamatsu, K., Yamada, N., Kanda, K., Haruyama, Y., and Matsui, S. 2006. Fluorinated diamond-like carbon coating as antisticking layer on nanoimprint mold. *Jpn. J. Appl. Phys.* **45**: L954–L956.

[9] Bailey, T., Choi, B.J., Colburn, M., Meissl, M., Shaya, S., Ekerdt, J.G. *et al.* 2000. Step and flash imprint lithography: Template surface treatment and defect analysis. *J. Vac. Sci. Technol. B* **18**: 3572–3577.

[10] Miki, K., Sakamoto, K., and Sakamoto, T. 1998. Surface preparation of Si substrates for epitaxial growth. *Surf. Sci.* **406**: 312–327.

[11] Fowkes, F.M. 1964. Attractive forces at interfaces. *Ind. Eng. Chem.* **56**(12): 40–52.

[12] Taniguchi, J., Machinaga, K., Unno, N., and Sakai, N. 2009. Filling behavior of UV nanoimprint resin observed by using a midair structure mold. *Microelectron. Eng.* **86**: 676.

[13] Delamarche, E., Bernard, A., Schmid, H., Bietsch, A., Michel, B., and Biebuyck, H. 1998. Microfluidic networks for chemical patterning of substrates: Design and application to bioassays. *J. Am. Chem. Soc.* **120**: 500.

[14] Fukuhara, M., Mizuno, J., Saito, M., Homma, T., and Shoji, S. 2007. Fabrication of metallic nanopatterns using the vacuum type UV-NIL equipment. *IEEJ Trans. Electr. Electron. Eng.* **2**(3): 307–312.

[15] Toyo Gosei Co., Ltd. 2009. Research & development (nanotechnology, bio). Available from: http://www.toyogosei.co.jp/eng/business/rd.html. [Accessed December 6, 2009.]

[16] Shibazaki, T., Shinohara, H., Hirasawa, T., Sakai, N., Taniguchi, J., Mizuno, J. *et al.* 2013. Desktop type equipment of thermal-assisted UV roller imprinting. *J. Photopolym. Sci. Technol.* **22**(6): 727–730.

[17] Shibazaki, T., Shinohara, H., Hirasawa, T., Sakai, N., Taniguchi, J., Mizuno, J. *et al.* 2009. Desktop type equipment of thermal-assisted UV roller imprinting. Proc. 20th Annual Meeting JSPP, June 3–4, Tokyo, Japan, pp. 327–328 (in Japanese).

[18] Perret, C., Gourgon, C., Lazzarion, F., Tallal, J., Landis, S., and Pelzer, R. 2004. Characterization of 8-in. wafers printed by nanoimprint lithography. *Microelectr. Eng.* **73/74**: 172–177.

[19] Hiroshima, H., Inoue, S., Kasahara, N., Taniguchi, J., Miyamoto, I., and Komuro, M. 2002. Uniformity in patterns imprinted using photo-curable liquid polymer. *Jpn. J. Appl. Phys.* **41**(6B): 4173–4177.

[20] Youn, S.W., Ogiwara, M., Goto, H., Takahashi, M., and Maeda, R. 2008. Prototype development of a roller imprint system and its application to large area polymer replication for a microstructured optical device. *J. Mater. Process. Technol.* **202**(1–3): 76–85.

[21] Hwang, S.Y., Hong, S.H., Jung, H.Y., and Lee, H. 2009. Fabrication of roll imprint stamp for continuous UV roll imprinting process. *Microelectr. Eng.* **86**(4–6): 642–645.

[22] Haisma, J., Verheijen, M., van den Heuvel, K., and van den Berg, J. 1996. *J. Vac. Sci. Technol. B* **14**: 4124.

[23] Kondo, M., Yasuda, H., and Kubodera, K. 1979. JP Patent S54-22389.

[24] Colburn, M., Suez, I., Choi, B.J., Meissl, M., Bailey, T., Sreenivasan, T.V. *et al.* 2001. *J. Vac. Sci. Technol. B* **19**(6): 2685.

[25] Toyo Gosei Co., Ltd. Business guide: Research & development. Available at: http://www.toyogosei.co.jp/eng/business/rd.html#nano_tech. [Accessed February 1, 2010.]

[26] Xu, F., Stacey, N., Watts, M., Truskett, V., McMackin, I., Choi, J. *et al.* 2004. Proc. SPIE, Vol. **5374**, p. 232.

[27] Sakai, N., Taniguchi, J., Kawaguchi, K., Ohtaguchi, M., and Hirasawa, T. 2005. *J. Photopolym. Sci. Technol.* **18**(4): 531.

[28] Ito, H., Houle, F.A., and Dipetro, R.A. 2006. Proc. SPIE, Vol. **6153**.

[29] Ito, H. 2007. *J. Photopolym. Sci. Technol.* **20**(3): 319.

[30] Kim, E.K., Stacey, N.A., Smith, B.J., Dickey, M.D., Johnson, S.C., Trinque, B.C., and Willson, C.G. 2004. *J. Vac. Sci. Technol., B* **22**: 131.

[31] Iyoshi, S., Miyake, H., Nakamatsu, K., Matsui, S. 2008. *J. Photopolym. Sci. Technol.* **21**(4): 573.
[32] Hagberg, E.C., Malkoch, M., Ling, Y., Hawker, C.J., and Carter, K.R. 2007. *Nano Lett.* **7**(2): 233.
[33] Kawaguchi, Y., Nonaka, F., and Sanada, Y. 2007. *Microelectr. Eng.* **84**: 973.
[34] Long, B. K., Keitz, B.K., and Willson, C.G. 2007. *J. Mater. Chem.* **17**: 3575.
[35] Kawaguchi, Y., Nonaka, F., and Sanada, Y. 2007. *Microelectr. Eng.* **84**: 973.
[36] Haatainen, T., Makela, T., Ahopelto, J., and Kawaguchi, Y. 2007. *Microelectr. Eng.* **86**: 2293.
[37] Sogo, K., Nakajima, M., Kawata, H., and Hirai, Y. 2007. *Microelectr. Eng.* **84**: 909.
[38] Schmitt, H., Zeidler, M., Rommel, M., Bauer, A.J., and Ryssel, H. 2008. *Microelectr. Eng.* **85**: 897.
[39] Schmitt, H., Frey, L., Ryssel, H., Rommel, M., and Lehrer, C. 2007. *J. Vac. Sci. Technol. B* **25**: 785.
[40] Tsunozaki, K. and Kawaguchi, Y. 2009. *Microelectr. Eng.* **86**: 694.
[41] Haering, E. 1984. In Parfitt, G.D. and Patsis, A.V. (eds), *Organic Coatings*. New York: Marcel Dekker, p. 79.
[42] Long, B.K., Keitz, B.K., and Willson, G. 2007. *J. Mater. Chem.* **17**: 3565.
[43] Hiroshima, H., Komuro, M., Kurashima, Y., Kim, S., and Muneishi, T. 2004. *Jpn. J. Appl. Phys.* **43**: 4012.
[44] Zhang, W. and Chou, S. 2003. *Appl. Phys. Lett.* **83**: 1632.
[45] Crivello, J.V. and Varlemann, U. 1995. *J. Polym. Sci. A* **33**(14): 2473.
[46] Crivello, J.V. and Lohden, G. 1996. *J. Polym. Sci. A* **34**(10): 2051.
[47] Crivello, J.V. and Sasaki, H. 1993. *J. Macromol. Sci. A* **30**(2&3): 189.
[48] Brady Jr,, R.F. 1992. *J.M.S. Rev. Macromol. Chem. Phys. C* **32**: 135.
[49] Yamaoka, T., Watanabe, H., Takahara, S., and Miyagawa, N. 1999. *Polym. Mater. Sci. Eng.* **81**: 55.
[50] Iyoshi, S., Miyake, H., Nakamatsu, K., and Matsui, S. 2008. *J. Photopolym. Sci. Technol.* **21**: 573.
[51] Miyake, H., Takai, H., Yukawa, T., and Iyoshi, S. 2009. *RadTech Jpn News Lett. Int. Ed.* **4**: 2.
[52] Owen, M.J. and Williams, D.E. 1991. *J. Adhes. Sci. Technol.* **5**: 307.
[53] Yoshino, N., Sato, T., Miyao, K., Furukawa, M., and Kondo, Y. 2006. *J. Fluorine Chem.* **127**: 1058.
[54] Yoshino, N. Patent JP 2004–107274.
[55] Yoshino, N. 2007. *Oleo Sci.* **7**: 513.
[56] Taniguchi, J., Kawasaki, T., Tokano, Y., Kogo, Y., Miyamoto, I., Komuro, M. *et al.* 2002. *Jpn. J. Appl. Phys.* **41**: 4194.
[57] Bailey, T., Choi, B.J., Colburn, M., Grot, A., Meissl, M., Shaya, S. *et al.* 2000. *J. Vac. Sci. Technol. B* **18**(6): 3572.
[58] Hirai, Y., Yoshida, S., Okamoto, A., Tanaka, Y., Endo, M., Irie, S. *et al.* 2001. *J. Photopolym. Sci. Technol.* **14**: 457–462.
[59] Sakai, N., Taniguchi, J., Kawaguchi, K., Ohtaguchi, M., and Hirasawa, T. 2005. *J. Photopolym. Sci. Technol.* **18**(4): 531.
[60] Chan, E.P. and Crosby, A.J. 2006. *J. Vac. Sci. Technol. B* **24**(6): 2716.

[61] Toyo Gosei Co., Ltd. Business guide: Research & development. Available at: http://www.toyogosei.co.jp/eng/business/rd.html#nano_tech. [Accessed February 1, 2010.]

[62] Houle, F.A., Miller, D.C., Fornof, A., Trung, H., Raoux, S., Sooriyakumaran, R. *et al*. 2008. *J. Photopolym. Sci. Technol.* **21**(4): 563.

[63] Tada, Y., Yoshida, H., and Miyauchi, A. 2007. *J. Photopolym. Sci. Technol.* **20**(4): 545.

[64] Schmitt, H., Frey, L., Ryssel, H., Rommel, M., and Lehrer, C. 2007. *J. Vac. Sci. Technol. B* **25**(3): 785.

[65] Houle, F.A., Fornof, A., Miller, D.C., Raoux, S., Truong, H., Simonyi, E. *et al*. Proc. SPIE, Vol. **6921**.

[66] Kim, E.K., Stacey, N.A., Smith, B.J., Dickey, M.D., Johnson, S.C., Trinque, B.C., and Willson, C.G. 2004. *J. Vac. Sci. Technol. B* **22**: 131.

[67] Sekiguchi, A., Kono, Y., and Hirai, Y. 2005. *J. Photopolym. Sci. Technol.* **18**(4): 543.

[68] Hua, F., Sun, Y., Gaur, A., Meitl, M.A., Bilhaut, L., Rotkina, L. *et al*. 2004. *Nano Lett.* **4**(12): 2476.

[69] Hua, F., Gaur, A., Sun, Y., Word, M., Jin, N., Adesida, I. *et al*. 2006. *Nanotechnology* **5**(3): 301.

[70] Johnson, S., Burns, R., Kim, E.K., Schmid, G., Dicky, M., Meiring, J. *et al*. 2004. *J. Photopolym. Sci. Technol.* **17**(3): 417.

[71] Hiroshima, H. 2005. *J. Photopolym. Sci. Technol.* **18**(4): 537.

[72] Colburn, M., Suez, I., Choi, B.J., Meissl, M., Bailey, T., Sreenivasan, S.V. *et al*. 2001. *J. Vac. Sci. Technol. B* **19**(6): 2685.

[73] Brooks, C.B., LaBrake, D.L., and Khusnatdinov, N. 2008. Proc. SPIE, Vol. **6921**.

[74] Katayama, J., Yamaki, S., Mitsuyama, M., and Hanabata, M. 2006. *J. Photopolym. Sci. Technol.* **19**(3): 397.

[75] Houle, F.A., Fornof, A., Sooriyakumaran, R., Truong, H., Miller, D.C., Sanchez, M.I. *et al*. 2007. Proc. SPIE, Vol. **6519**.

[76] Ahn, S.H., Kim, J., and Guo, L.J. 2007. *J. Vac. Sci. Technol. B* **25**(6): 2388.

[77] Ogino, M., Kaji, M., and Rai, H. 2007. JP Patent 2007–186570 (unexamined).

[78] Fukuhara, M., Ono, H., Hirasawa, T., Ohtaguchi, M., Sakai, N., Mizuno, J., and Shoji, S. 2007. *J. Photopolym. Sci. Technol.* **20**(4): 549.

[79] Willson, C.G. and Trinque, B.C. 2003. *J. Photopolym. Sci. Technol.* **16**(4): 621.

[80] Yoneda, I., Mikami, S., Ota, T., Koshiba, T., Ito, M., Nakasugi, T., and Hisashiki, T. 2008. Proc. SPIE, Vol. **6921**.

[81] Shinohara, H., Fukuhara, M., Hirasawa, T., Mizuno, J. and Shoji, S. 2008. *J. Photopolym. Sci. Technol.* **21**(4): 591.

6

Applications and Leading-Edge Technology

Jun Taniguchi[a], Hidetoshi Shinohara[b], Jun Mizuno[c], Mitsunori Kokubo[d], Kazutoshi Yakemoto[e], and Hiroshi Ito[e]

[a]*Department of Applied Electronics, Tokyo University of Science, Japan*
[b]*Department of Electronic and Photonic Systems, Waseda University, Japan*
[c]*Nanotechnology Research Laboratory, Waseda University, Japan*
[d]*Toshiba Machine Co., Ltd, Japan*
[e]*The Japan Steel Works, Ltd, Japan*

6.1 Advanced Nanoimprinting Technologies

Advanced nanoimprinting technologies are introduced in this chapter. Worldwide trends in nanoimprinting technologies and topics related to them are described. The resolution limit of nanoimprint lithography is mentioned, improved nanoimprinting technologies are introduced, and roll-to-roll nanoimprinting technologies are described.

Nanoimprint Technology: Nanotransfer for Thermoplastic and Photocurable Polymers, First Edition. Edited by Jun Taniguchi, Hiroshi Ito, Jun Mizuno, and Takushi Saito.

6.1.1 Resolution Limit of Nanoimprint Lithography

We are interested in how small patterns can be fabricated using nanoimprinting technology. The resolution limit of NIL depends on the fineness of the molds used, and the molds are usually fabricated using electron beam lithography. However, it is difficult to use EBL in order to delineate stripe patterns consisting of lines and spaces that are both less than 15 nm wide. Therefore, other methods of delineating such stripe patterns have previously been attempted by several groups. Full pitch (i.e., line and space widths together) and half pitch (i.e., only line width) are important parameters for producing semiconductors because these parameters determine the degree of ultra-large-scale integration in semiconductor circuits. Thus, the resolution limits of the full-pitch and isolated patterns are introduced.

Because it is difficult to use EBL to fabricate fine-pitch patterns, Chou and his group used molecular beam epitaxy (MBE) and selective wet etching to fabricate the mold [1]. Figure 6.1 shows a schematic of the method they used to fabricate the mold.

A GaAs/$Al_{0.7}Ga_{0.3}As$ superlattice substrate is first grown using MBE (Figure 6.1(a)). MBE can be used to precisely control the thickness of each layer; thus, it is possible to control the degree of fineness of the pitch. After the superlattice substrate is grown using MBE, it is mechanically broken to form a cleaved edge (Figure 6.1(b)). The cleaved edge is immersed in a dilute solution of hydrofluoric acid (HF) (Figure 6.1(c)) to selectively etch and remove the $Al_{0.7}Ga_{0.3}As$. After the cleaved edge is etched, it is used to produce a fine-pitch pattern mold (Figure 6.1(d)). A 14 nm-wide pitch consisting of 7 nm-wide lines and spaces was then obtained using this mold and UV nanoimprinting. This method was first reported in 2004 and was used at that time to produce a fine-pitch mold that held the world record until recently for the finest pitch mold ever produced. However, helium ion beam lithography was recently used to produce an even finer pitch mold; thus breaking the world record.

Recent advances in scanning helium ion microscope technology have enabled helium ion beams to be focused on spots less than 0.24 nm in diameter, which is suitable for lithography. In addition, the mass of a helium ion is 7000 times greater than that of an electron; therefore, a helium ion beam causes less forward scattering in resist materials than an electron beam. Therefore, helium ion beam lithography can be used to delineate fine-pitch line and space patterns. In 2012, Hewlett-Packard Labs produced an 8 nm-wide full-pitch pattern mold that was delineated using helium ion beam

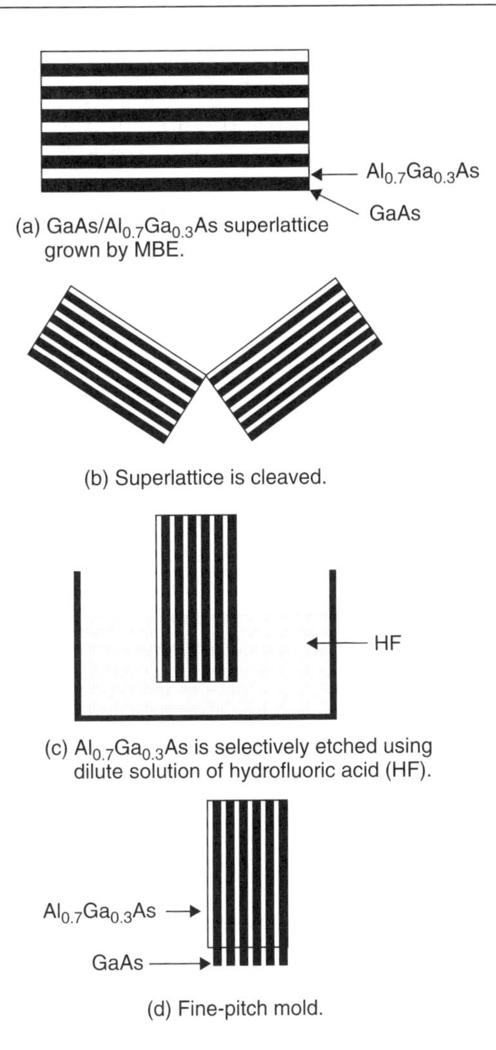

Figure 6.1 Schematic of the method used by Chou *et al.* to fabricate a fine-pitch mold

lithography for hydrogen silsesquioxane resists and produced an 8 nm-wide full-pitch pattern (i.e., a 4 nm-wide line and space) that was imprinted using UV nanoimprinting technology [2]. It is the smallest pitch pattern ever produced.

A 2.4 nm-wide line, in contrast, was obtained for the resolution limit of the isolated pattern produced by Rogers and his group [3]. They used carbon nanotubes and UV nanoimprinting to fabricate the isolated pattern. It is therefore possible to use nanoimprinting technology in order to fabricate single nanoscale-line patterns. Current applications of such fine-pitch patterns include ULSI and bit-patterned media for hard disk drives. Other applications are also desirable.

6.1.2 Improved Nanoimprinting Technologies

Repeating semiconductor processes produces semiconductor substrates that exhibit uneven surfaces and compound semiconductor substrates such as GaN and GaAs substrates that exhibit severely warped and undulated surfaces (Figure 6.2(a)). A hard stamp cannot be used to conformably imprint the surfaces of such substrates (Figure 6.2(b)). Imprinting is used to produce trench patterns consisting of different residual layers on the surface of the substrates (Figure 6.2(c)). The trench patterns are used as masks during dry etching, and some patterns are destroyed during deep etching. A method in which soft molds and air pressure are combined has recently been developed to solve this problem [4]. Soft molds are first prepared using either thermal or UV nanoimprinting (Figure 6.3(a)). The soft molds are then set onto substrates that have undulated surfaces, and the molds and substrates are

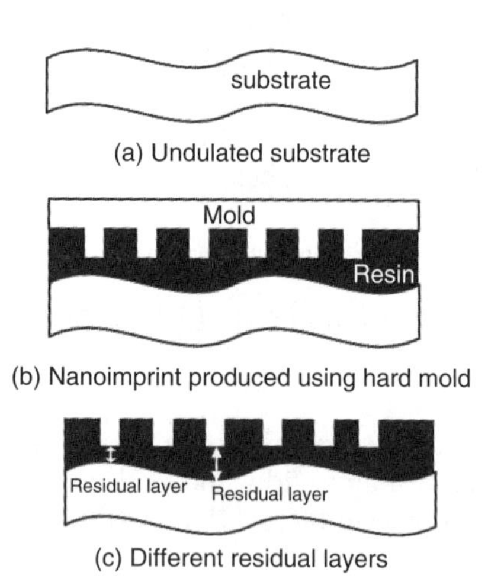

Figure 6.2 Nanoimprinting using flat mold on substrate whose surface is undulated

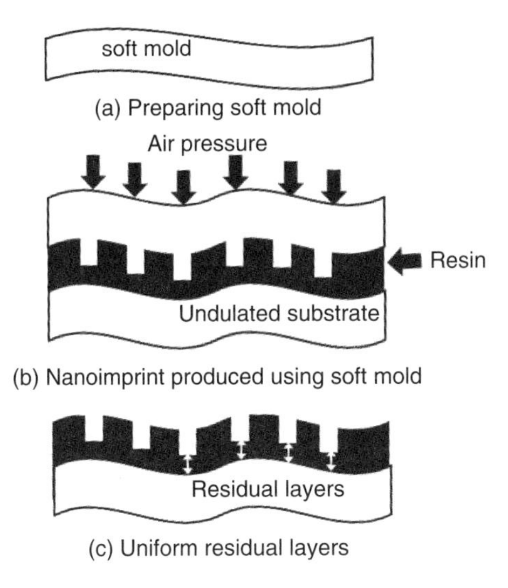

Figure 6.3 Nanoimprinting combining soft mold and air pressure on substrate whose surface is undulated

pressed together with air pressure. When air pressure and soft molds are combined, the patterns produced on the soft molds are deformed on the basis of the shape of the undulated substrates (Figure 6.3(b)). Nanopatterns are then fabricated on the undulated substrates such that the nanopatterns exhibit uniform heights and/or residual layers (Figure 6.3(c)). Thus, dry etching can be used to fabricate uniform nanopatterns on substrates whose surfaces are either warped or undulated. This combined process is very useful for manufacturing light-emitting diodes (LEDs), which require patterning on the surfaces of GaN substrates in order to enhance light emission. In addition, this combined process can be used to uniformly transfer nanopatterns along whole 6 to 8 inch wafers.

Flexible molds such as those produced from poly(dimethylsiloxane) (PDMS) can be used to transfer nanopatterns onto curved surfaces. Flexible molds containing micropatterns or nanopatterns such as micro lenses or moth-eye structures are contacted onto curved, warped, or undulated surfaces, and the patterns are transferred using either UV-NIL or contact printing [5]. A mixture of micropatterns and nanopatterns can then be obtained on curved substrates. Combining flexible molds with either UV-NIL or contact printing is useful for producing optical devices.

Nanopatterns are also very useful for application in biotechnologies. For example, nanopatterned surfaces can be used as cell culture plates.

Nanopattern sheets act as scaffolds upon which the cells can grow [6]. In addition, nanopatterns can enhance antigen–antibody reactions. It is possible to use NIL in order to fabricate nanopattern plates and sheets at low cost and high throughput.

6.1.3 Roll-to-Roll Nanoimprinting Technologies

High-throughput and high-resolution NIL is a powerful method of fabricating next-generation devices. RTR imprinting is of particular interest [7] because it delivers very high throughput. Since the first reports on RTR-NIL by Chou and his group [8], research has expanded to include techniques of transferring patterns, processes of fabricating roll molds, and various new methods of RTR-NIL. For instance, the VTT (Technical Research Centre of Finland) group has been studying thermal RTR-NIL for a long time, and they recently used it to achieve a feed speed as high as 20 m/min [9]. They wrapped a roll mold in nickel-electroplated foil and used cellulose acetate as a transfer web. RTR UV-NIL, in contrast, has been used to achieve a feed speed of 18 m/min. A 100 nm-wide, 200 nm-deep line-and-space pattern was obtained using a replica mold and high-intensity UV-LED illumination [10].

The main problem associated with RTR-NIL is that fabricating roll molds is difficult. However, replica mold materials that exhibit good release properties and high durability have recently been developed. Materials that are currently used to produce replica molds include polytetrafluoroethylene (PTFE) [11], ethylene tetrafluoroethylene (ETFE) [12], and polyurethane acrylate (PUA) [3].

However, it is difficult to fabricate cylindrical roll molds for RTR-NIL. Roll molds are usually fabricated by rolling up nickel-plated foil that has previously been imprinted with a nanopattern. However, this action generates a seam. A method in which an electron beam is used to directly write a nanopattern onto a rotating cylindrical substrate in order to produce a seamless nanoscale mold has recently been developed [13]. The method involves using a cylindrical substrate (i.e., a roll-mold substrate) that has previously been coated with a resist. The EB then directly writes the desired nanopattern onto the resist-coated substrate while the substrate is rotated in a vacuum. However, the throughput associated with this method is very low because EB writing requires a long time to delineate the nanopattern. A method that involves directly transferring a nanopattern from a small mold containing the pattern to increase the diameter of the large roll has recently been reported in an effort to solve this problem [14]. The roll mold was produced using synthetic quartz, and the diameter of the roll mold was 30 mm. The roll mold was dipped into hydrogen silsesquioxane (HSQ), which acts

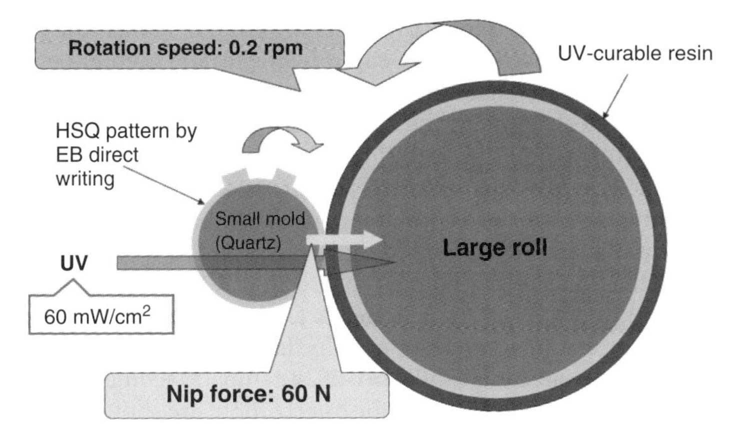

Figure 6.4 Schematic of method used to increase the diameter of a large roll by directly transferring the line pattern from a small patterned roll to the large roll

as a high-resolution negative-type EB resist. The EB resist-coated mold was subjected to EB lithography. After the EB lithography was complete, 520 nm-wide line patterns were obtained on the quartz roll mold. The transparent roll mold was subsequently used to directly transfer the line patterns onto a large mold rotated at 0.2 rpm under a nip force of 60 N and at a UV dose of 60 mW/cm^2 (Figure 6.4). The diameter of the large mold was 150 mm, and the mold was covered with a UV-curable film. The average width of the transferred lines was 510 nm. RTR UV-NIL was carried out using a large mold made of the resulting patterned polymer, and the line patterns were transferred from the small mold to the large one. The method of increasing the diameter of the large roll is therefore very effective for performing RTR UV-NIL because a small-diameter roll can be used as the master mold.

6.2 Applications

6.2.1 Seamless Pattern[1]

One of the widely used methods for the fabrication of large-area patterns is step-and-repeat imprint lithography [16–18]. Continuous patterns can be

[1] The work in this section was conducted jointly by Waseda University and Toshiba Machine Co., Ltd [15].

fabricated by the step-and-repeat process. However, in reality, the photocurable resin around the mold is partly cured by the leaked UV light even in the case of drop-on-demand printing. Replication errors occur because the residual cured resin interferes with the subsequent step in imprint lithography. To solve this problem, a litho/etch/litho/etch double-patterning technique was applied to the UV-NIL process. The proposed UV-NIL process with multiple etching steps is called "double UV-NIL."

The principle of the proposed double-pattening UV-NIL is shown in Figure 6.5. In the first step, a photocurable resin (TR-21, Toyo Gosei Co., Ltd) was spin-coated on a quartz substrate. A Cr layer was previously coated on the surface of the substrate as a transfer layer. Replicated patterns were formed using the resin by UV-NIL (ST-50, Toshiba Machine Co., Ltd). The dimensions and a magnified view of the quartz mold are shown in Figure 6.6. The mold had a line-and-space pattern with width of 100 nm, height of 100 nm, and pitch of 150 nm. Next, the residual resin layer was etched off by O_2 plasma reactive ion etching (O_2-RIE). The Cr layer was patterned by wet etching using cerium ammonium nitrate. Finally, the remaining resin layer was removed completely by wet etching.

In the second step, the photocurable resin was again coated on the wafer. To overlap this mold in an appropriate area on the previously coated chrome pattern, the mold and wafer were aligned accurately. Then, the second replicated pattern was formed. Next, the residual resin was removed and Cr was etched again by the same method. The Cr patterns used for quartz etching were formed on the entire wafer.

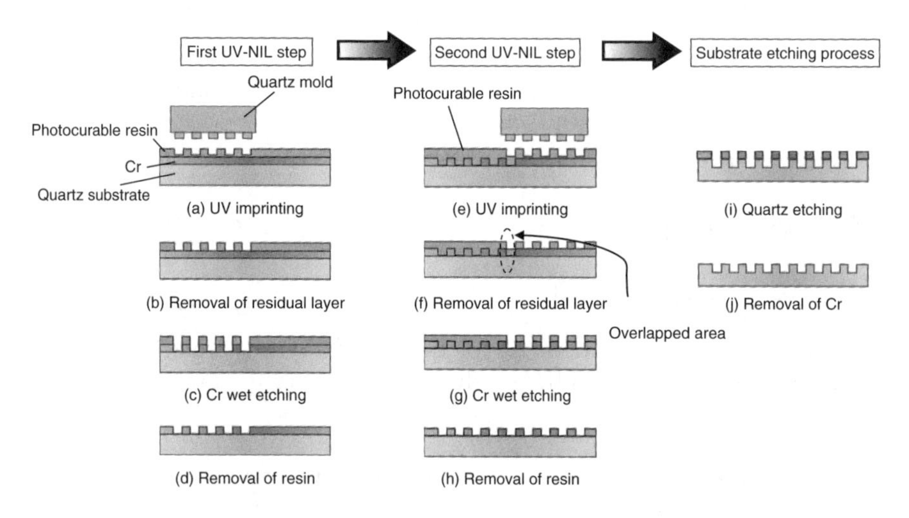

Figure 6.5 Double-patterning UV-NIL process

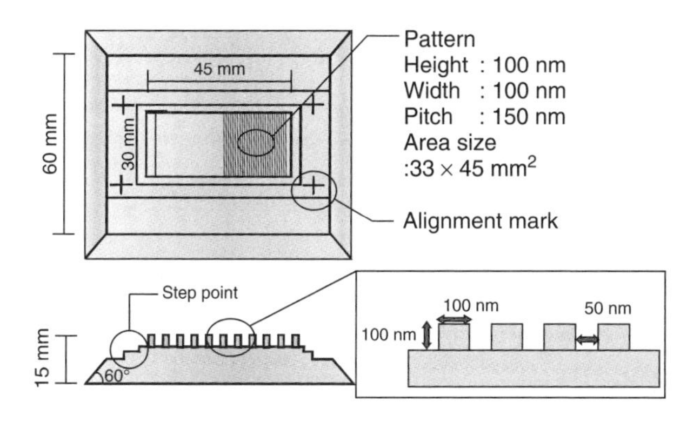

Figure 6.6 Dimensions and magnified view of quartz mold

In the substrate etching process, the quartz wafer was etched by ICP-RIE. The Cr layer was used as a hard mask for transferring the nanopattern onto the quartz wafer. After ICP-RIE, the remaining Cr was removed from the quartz wafer. A large-area seamless quartz pattern was formed by this proposed method.

A SEM image of the etched quartz surface is shown in Figure 6.7. The quartz pattern had a height of 197 nm, width of 50 nm, and pitch of 150 nm. These results indicate that quartz patterns can be formed successfully.

The results of the double-patterning UV imprinting process are shown in Figure 6.8. To accurately evaluate the pattern accuracy, SEM images of three parts of the patterns were magnified, as shown in Figure 6.9. In the case of the mold pattern with a width of about 50 nm, the line-width errors for the three parts were within 10 nm. These images show that seamless patterns were fabricated by the proposed double-patterning UV imprinting process. This result indicates that the seamless large-area nanopattern could be fabricated by step-and-repeat imprint lithography. Thus, the proposed method is useful for forming large-area patterns with high throughput and low cost.

6.2.2 Multistep Cu Interconnection[2]

Advanced LSI circuits and smart sensor systems require nanoscale interconnections to transmit signals. In recent years, multilayer interconnection technology has been developed as the main technology for internal wiring

[2] The work in this section was conducted jointly by Waseda University and Toppan Printing Co., Ltd [19].

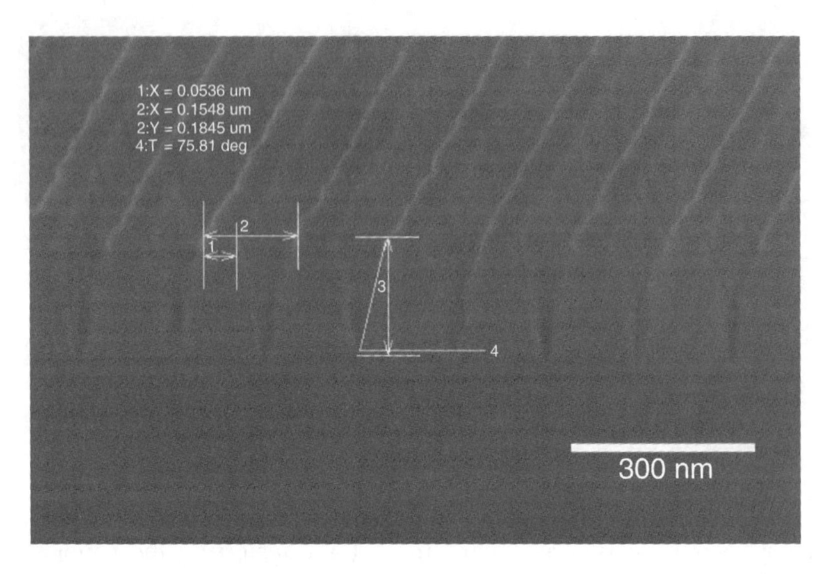

Figure 6.7 Cross-sectional SEM image of quartz mold after ICP-RIE: pillar height, 197 nm; width, 50 nm; pitch, 150 nm

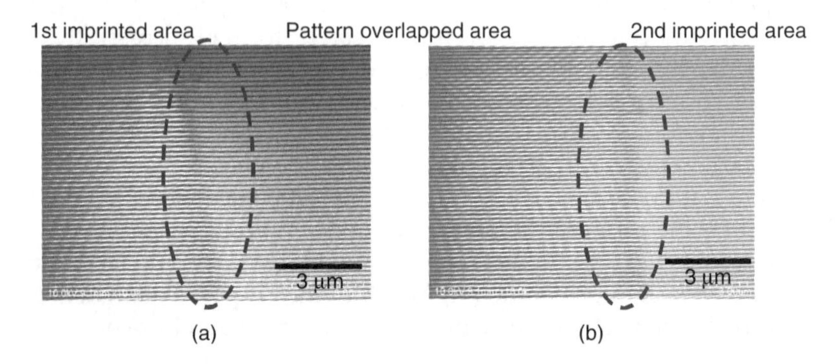

Figure 6.8 SEM image of seamless pattern on quartz wafer: (a) after first UV-NIL step, including overlapped area and (b) after second UV-NIL step, including overlapped area (dashed ellipses indicate interfaces between area after first or second UV-NIL step and overlapped area)

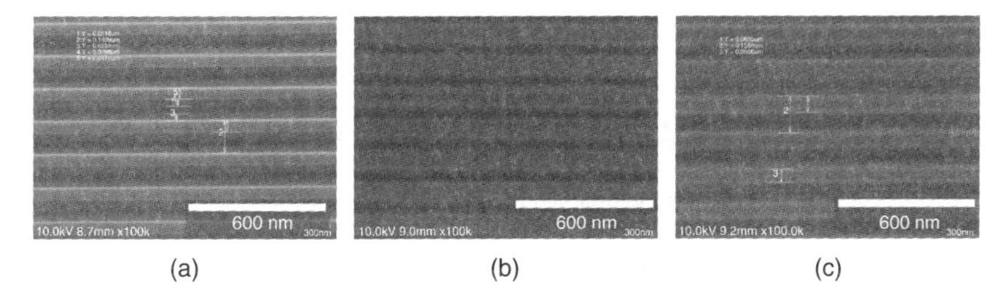

Figure 6.9 SEM images of fabricated patterns: (a) after first UV-NIL step, (b) overlapped area, (c) after second UV-NIL step

and forming connections in electronic devices. Since the conventional multilayer interconnections for integrated circuits use via holes that increase the dead space on the board, the actual wiring space decreases. This problem is solved by employing the build-up approach in which metal lead layers and other layers are stacked one by one on the core substrate. Since this approach retains large space on the board, high-density multilayer interconnections are realized. However, the problem with this approach is that stacking many layers requires a complex fabrication process. A conventional process requires separate lithography and etching steps to pattern vias and trenches. Using multistep molding for UV-NIL, patterns consisting of via holes and trench steps were formed simultaneously [20].

The fabrication process of a multistep interconnection by the UV-NIL technology is shown in Figure 6.10. First, a seed layer consisting of Cr and Cu was formed on a silicon dioxide substrate. Then, a 360 nm-thick photocurable resin (TR-21) was spin-coated on the seed layer. Via holes and trenches were simultaneously formed by UV imprinting using a mold made of quartz. The thin resin membrane on the seed layer was removed by O_2-RIE. Then, via and trench patterns were formed by Cu electroplating. Finally, excess electroplated Cu was flattened by chemical mechanical polishing (CMP).

Figure 6.11 shows a SEM image of the mold structure with the via hole and trench pattern. The mold shows a patterned array. A pattern consists of a trench with a size of 600×2000 nm^2 and a via with a size of 360×360 nm^2.

Figure 6.12 shows a cross-sectional SEM image of a multistep pattern after UV imprinting. The 600 nm-wide trench pattern and the 360 nm-wide via pattern were formed simultaneously. Figure 6.13 shows cross-sectional SEM images of the Cu interconnection structures. The Cu multistep structures were observed on the seed layer. There was no void between the Cu multistep structures and the resin. The Cu structure also had no voids, indicating that it could be used for forming electrical connections. The Cu structures showed a 600 nm-wide trench and 350 nm-wide via. Since the photocurable

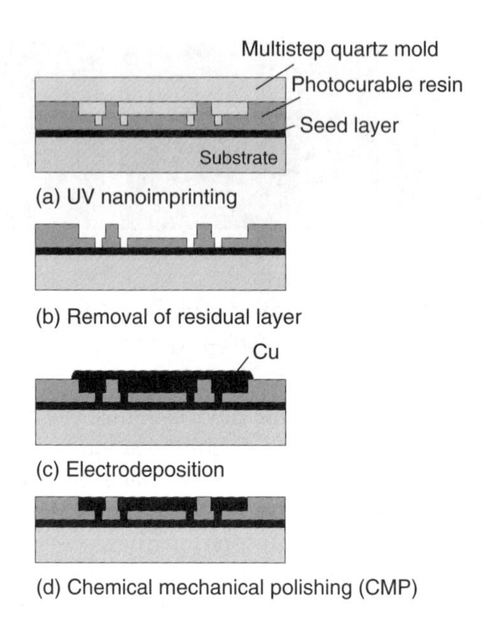

Figure 6.10 Fabrication process of multistep Cu interconnection by UV-NIL

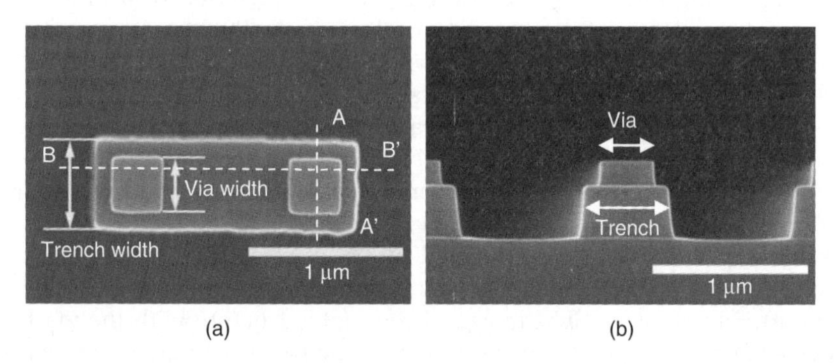

Figure 6.11 SEM image of mold structure with via hole and trench pattern: (a) top view, (b) cross-sectional view

resin had high durability against electrodeposition, the imprinted structures obtained before and after Cu electrodeposition exhibited negligible changes. Figure 6.14 shows top-view SEM images of Cu multistep structures before CMP. This figure shows that metallic structures were fabricated individually and almost uniformly. The SEM images of the Cu patterns after CMP are

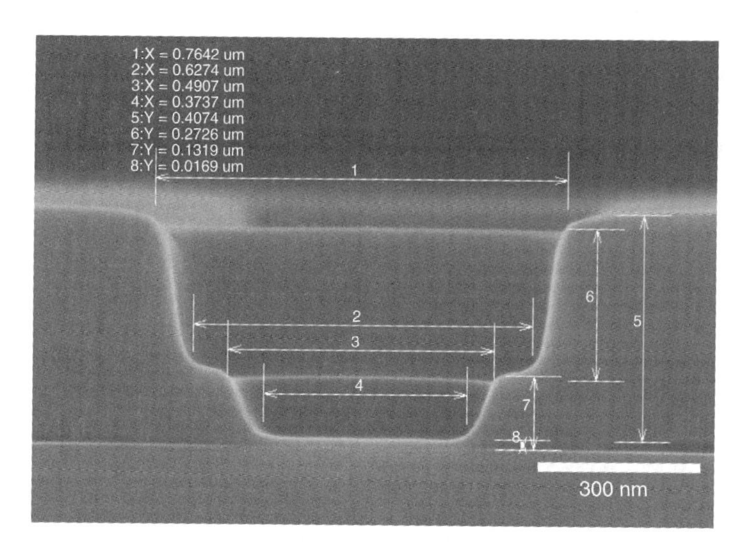

Figure 6.12 Cross-sectional SEM image (corresponding to line A–A′ in Figure 4.7) of multistep patterns after UV imprinting

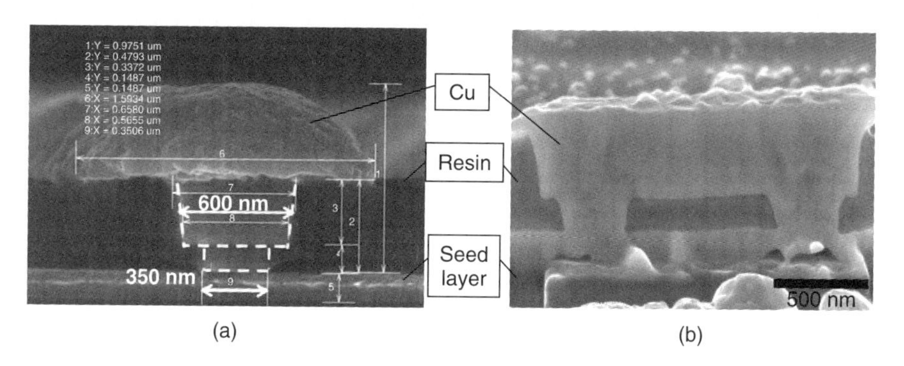

Figure 6.13 Cross-sectional SEM image (corresponding to (a) line A–A′, (b) line B–B′ in Figure 6.11) of Cu interconnection structures before CMP

shown in Figure 6.15. This image indicates that extra electrodeposited Cu was successfully removed and planarized after CMP.

This multistep molding process is simple and effective. In the near future, it is planned to determine the electrical characteristics of the Cu double-deck structure. By using a low *k*-type photocurable resin, it is hoped to directly fabricate multistep patterns in an insulating layer by this method and skip

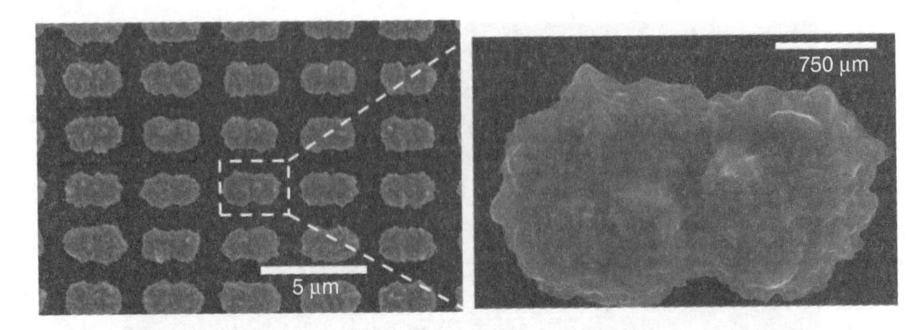

Figure 6.14 Top-view SEM images of Cu multistep structures before CMP

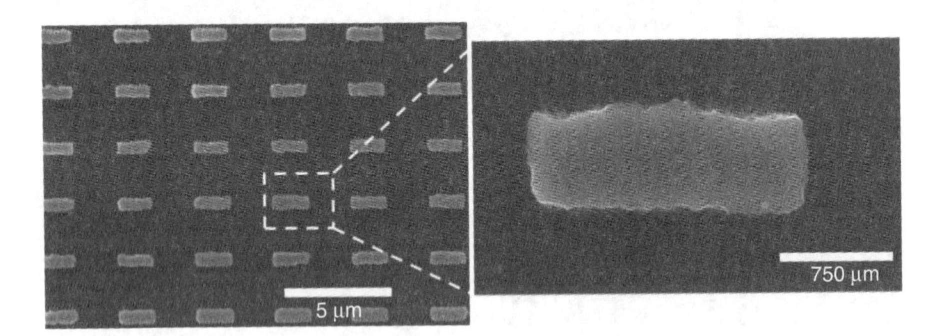

Figure 6.15 Top-view SEM images of Cu multistep structures after CMP

the etching process. In addition, this method is expected to be applicable for designing high-density interconnections desired in advanced LSI circuits and high-performance smart sensor systems.

6.2.3 GaN Nanostructures for High-Intensity LED[3]

Light-emitting diodes are expected to be used as interior illuminations, car lightings, and displays because of their low maintenance cost and low carrying charges. However, currently, the luminescence efficiency of LEDs is lower than that of a conventional fluorescent lamp. If the luminescence efficiency is increased, LEDs can be applied widely to common lighting systems.

[3] The work in this section was conducted jointly by Waseda University and Sumitomo Chemical Co., Ltd [21].

High-intensity LEDs can be developed by improving the internal quantum and light extraction efficiencies. Light extraction efficiency is increased by reducing the reflection of light off the surface. In conventional LEDs, this efficiency is low because the majority of light reflects off the smooth top surface and is absorbed in the LED. To improve the light extraction efficiency, a rough surface that has random uneven structures and photonic crystal structures is useful. The top rough surface reduces internal light reflection and scatters the light outward [22]. A photonic crystal is an optical material whose refractive index changes periodically and causes light diffraction for a particular wavelength. Thus, the photonic crystal structures formed on the surface can be used for controlling the reflection of light. The diffraction grating can be used to extract the first-order light from the incident light over the critical angle [23]. Therefore, nanostructures are useful for fabricating anti-reflection materials [24–28]. A nanoscale metal mask was fabricated for dry etching over a large area by UV-NIL in combination with electrodeposition. The nanostructures of GaN for anti-reflection were fabricated by RIE using this mask.

Figure 6.16 shows the cross-section of an LED with nanostructures. Since nanoscale patterns were periodically formed with a size almost equal to the light wavelength of the LED, it was expected to behave as a photonic crystal. Multiple scattering by a photonic crystal results in the formation of photonic band gaps that prohibit the propagation of specific wavelengths.

The fabrication process of the LED is shown in Figure 6.17. First, a thin metal film to be used as a seed layer for electrodeposition was formed on the GaN substrate surface. In addition, a photocurable resin (TR-21) was spin-coated on the seed layer. Next, the replicated patterns on the photocurable resin were formed by UV-NIL (Figure 6.17(a)). This photocurable resin could tolerate the bath liquid used for Ni electrodeposition. The quartz mold for UV-NIL showed dots with a depth of 225 nm, diameter of 200 nm, and pitch of 500 nm. A pressure of about 1.3 MPa was applied to the mold. The residual layer was etched off by O_2-RIE. Next, Ni nanopatterns were electrodeposited

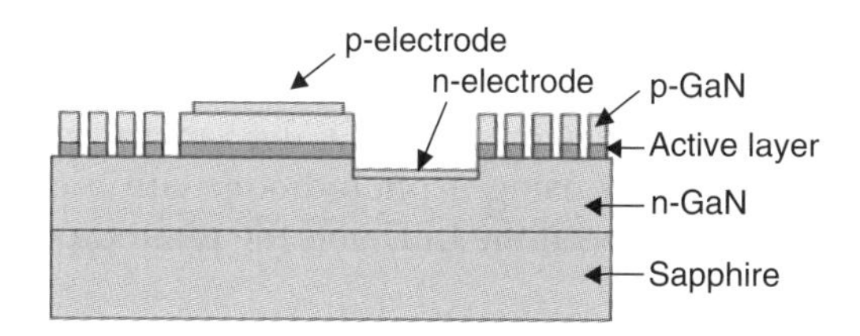

Figure 6.16 Cross-section of LED with nanostructures

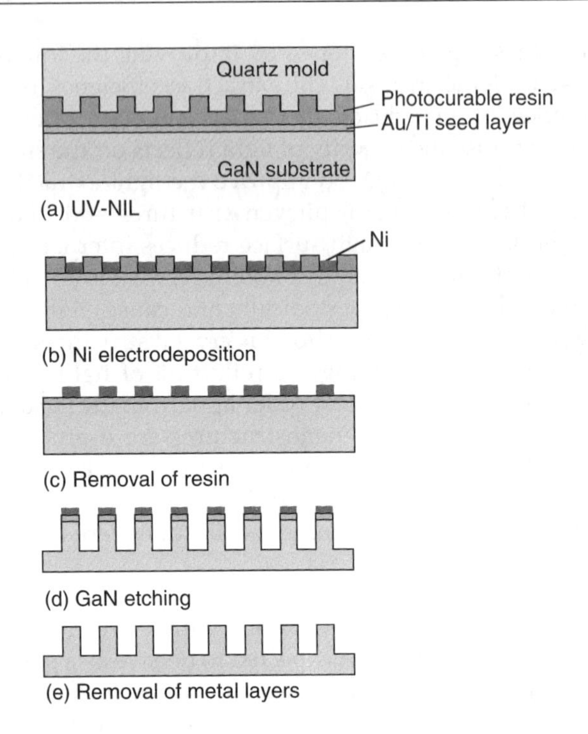

Figure 6.17 Fabrication process of GaN nanostructures

in the holes using the seed layer (Figure 6.17(b)). Ni was used as an etching mask because it is highly resistant to Cl_2 plasma. The photocurable resin was removed after electrodeposition (Figure 6.17(c)). Subsequently, the GaN substrate was etched by Cl_2-based ICP-RIE (Figure 6.17(d)). Finally, the Ni mask and seed layer were removed (Figure 6.17(e)). Nanostructures with a depth of 600 nm, diameter of 300 nm, and pitch of 500 nm were formed on the GaN surface. The luminescence wavelength of the LED was 474 nm. The pitch of the GaN structures was almost equal to the luminescence wavelength.

Figure 6.18 shows SEM images of the fabricated GaN nanostructures. The nanostructures were 600 nm in depth. It was confirmed that the Ni film was fairly resistant to dry etching.

After forming the electrodes on the LED, the radiant intensity from the front surface was measured using a photodiode with a detection area of 78.5 mm^2. The distance between the LED and the photodiode was 51.5 mm.

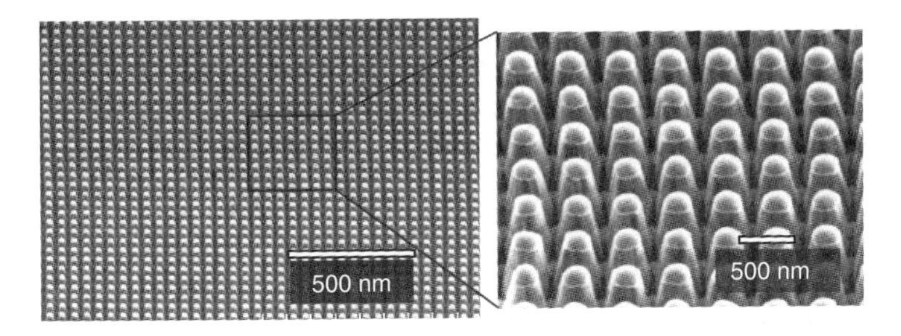

Figure 6.18 SEM images of fabricated GaN nanostructures

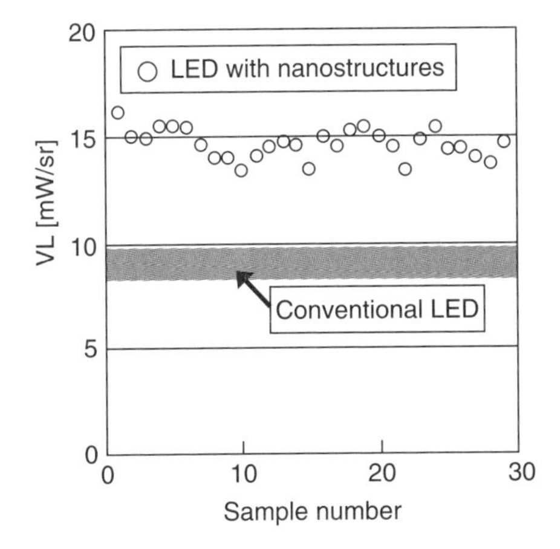

Figure 6.19 Radiant intensity of LED front surface

This setup was used to measure the intensity of the LED on a wafer. Figure 6.19 shows the measured radiant intensity of the fabricated LED. The horizontal axis indicates sample number. The vertical axis indicates radiant intensity (mW/sr). The straight line represents the radiant intensity of a conventional LED without the nanostructure. The circles denote the radiant intensity of the fabricated LED. The forward voltage of the LED was 4.8 V, which was the same as that of the conventional LED. The radiant intensity of the LED was 1.5 times larger than that of the conventional LED.

Since this method enables the fabrication of wafer-level GaN nanostructures at low cost, it is applicable to the fabrication of photonic crystals.

6.3 High-Accuracy Nanoimprint Technology, Development of Micropatterning Method, and Automatic Process Control Using Batch Press Type, Step and Repeat Type Nanoimprint Machine

6.3.1 Introduction

The nanoimprint technology process [29] is a method of forming a nanometer-level structure by pressing a mold with several tens of nanometers to several tens of micrometers micropattern to surfaces of resin, plastic, and glass material, etc. Because a UV nanoimprint method has been taken up in the ITRS (International Semiconductor Technological Road Map) as a candidate for "next generation semiconductor exposure process," it became a nanoprocessing method highlighted across the world.

In this technology, being able to achieve nanometer-level pattern formation at low cost is attractive; there is no expensive process device such as an optical exposure machine or the devices of electron beam drawing apparatus [30]. Now, the technologies of material, mold, and apparatus which compose nanoimprint process development have accelerated rapidly. It seems that the nanoimprint process has spread as a general nanoprocessing technology, and the application development will advance in the future. Here, the outline of the thermal imprint process and thermal imprint apparatus is explained.

6.3.2 Thermal Imprint

6.3.2.1 Thermal Imprint Process

The nanoimprint process is distinguished by the difference in energy used when molding by "thermal imprint" and "UV imprint." Recently, "room temperature imprint" which uses spin on glass (SOG) for the transcript material has joined them. It is one of the methods noted for its simple imprint mechanism and characteristic material (Figure 6.20). We call this method "COLD IMPRINT®."

Moreover, it is classified into (i) direct imprint, (ii) roller imprint, and (iii) R2R imprint by the difference in press mechanism (Figure 6.21). Original specifications and characteristics are proposed for each company target where case (i) is the target device. For instance, "step & repeat is possible," "imprinting with the decompression atmosphere is possible," and "a large area can be imprinted." The work of Toshiba Machine Co., Ltd on the thermal–direct imprint technique and the possible development of an automatic machine are explained here.

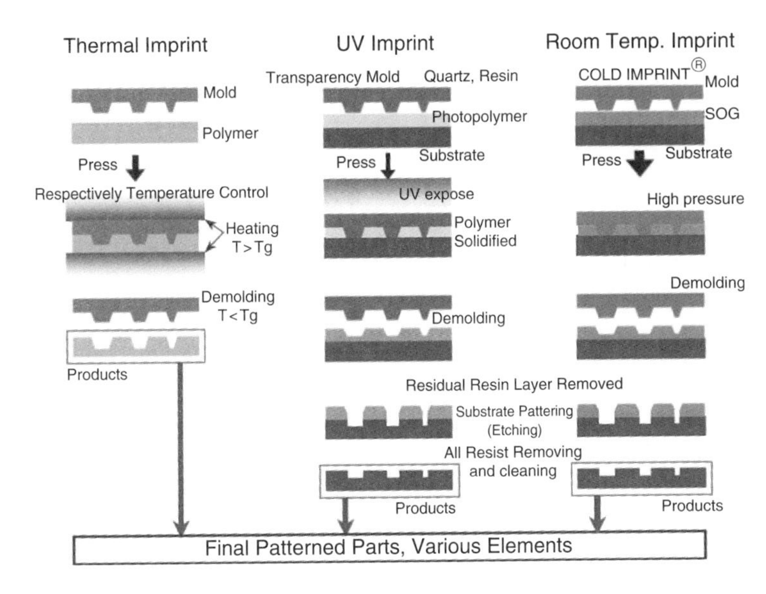

Figure 6.20 Process comparison between thermal, UV, and room temperature imprint

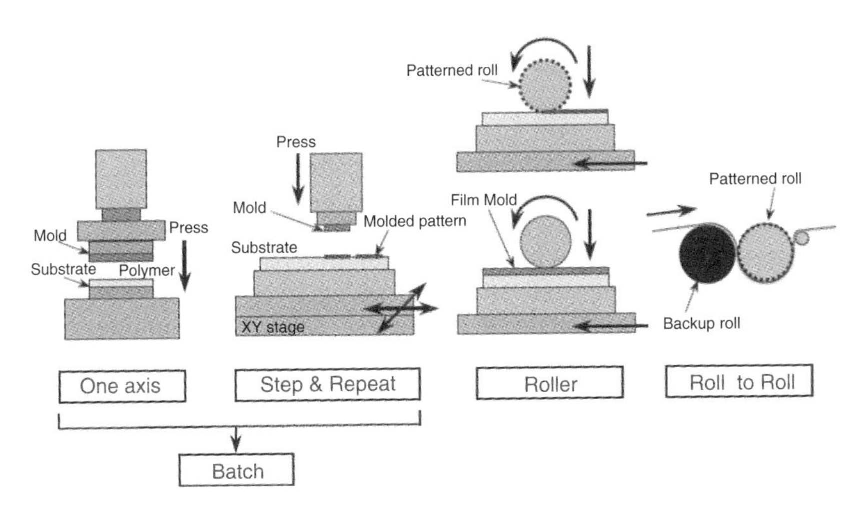

Figure 6.21 Several kinds of nanoimprint method

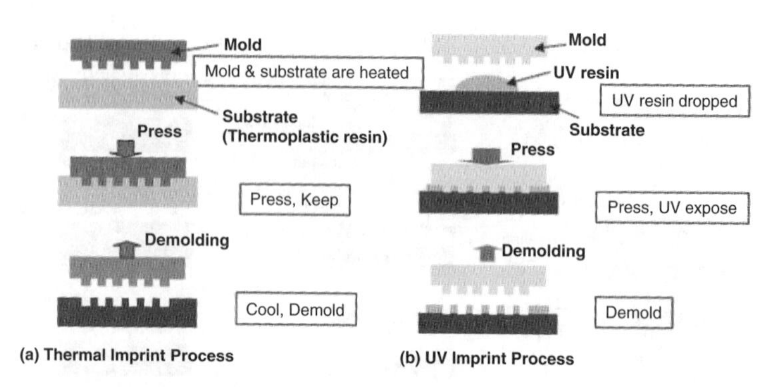

Figure 6.22 Flow of nanoimprint process

Figure 6.22 shows a basic process of thermal-direct imprint. The mold with micropattern of monocrystalline silicon, quartz glass, or nickel and the molded material (for instance, thermoplastic resin) are opposed in parallel. Next, the mold and resin are heated to more than the glass metastasis temperature of the resin. Then, the mold and material are pressed so that the pressure on the surface may become uniform with the temperature maintained. In addition, the temperature of the mold and resin are cooled below the glass metastasis temperature after maintenance of heating and pressurizing. As a result, the resin stiffens. Finally, the mold is demolded from the substrate. This thermal imprint has the characteristics that it can be applied to various thermoplastic resins and heat-stiffening resins, the selection leg of the resin corresponding to the usage is wide, and the selection leg of the type material is also wide. In this process, the transcript pattern is often used as it is a function of the structural body, and applied to a micro-optical element (micro lens, wavelength separation device, etc.) and bio device (micro passage, cell culture device, etc.).

Here, the basic composition of the thermal imprint is introduced. It is important to decide what type of imprint method you should use, according to the error factor which the substrate and mold possess. This is explained in Figure 6.23. First, a standard composition of the mold and substrate is shown in (A) and (B). (A) is a composition which uses "the mold with large stiffness" and presses with a highly accurate hard plane. (B) uses "the mold with small stiffness" and has "uniform load mechanism in-plane" uniformly pressed to the back of the mold. Though the flatness of the pattern side is excellent, the substrate has poor accuracy in terms of parallelism on the pattern side at the bottom (C). In this case, the mold base only has to provide a function to adjust the angle, and composition (A) can be used. The undulation of the

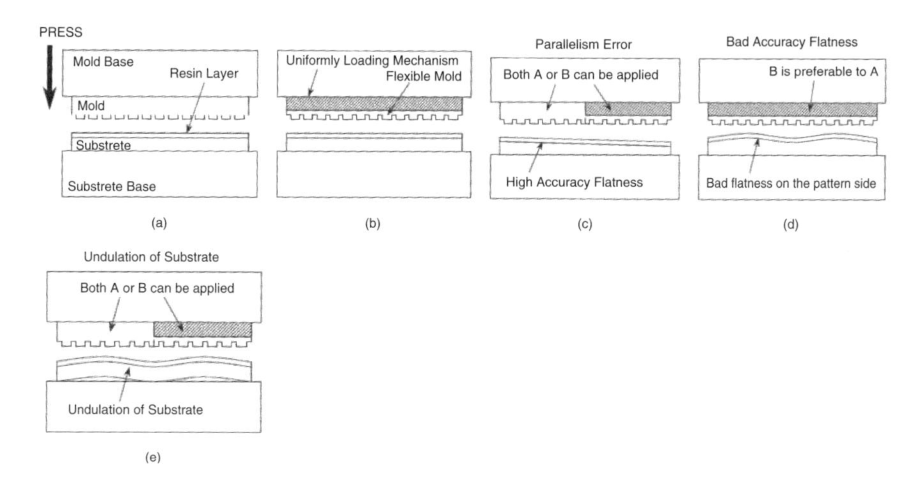

Figure 6.23 Several kinds of basic composition for nanoimprint process

substrate is corrected by applying a high load using the large stiffness mold. The flatness of the patterned surface is poor (D), but indispensable with the flexible mold and uniform load mechanism.

Thus, it is important in the following to select the imprint technology and the machine. "After the condition which cannot be changed is determined, the substrate, mold, and method of pressing can be chosen."

6.3.2.2 Thermal Imprint Machine

Toshiba Machine Co., Ltd develops and sells a machine that corresponds to thermal imprint, UV imprint, room temperature imprint, and micro contact imprint (soft imprint) [31]. In the standard model "ST50," the maximum press force is 50 kN. However, there are many results available for "ST200," "ST01," and "ST02." It is demanded that the imprint machine correspond flexibly to various materials and mold shapes, resin, substrate, and the imprint method. Therefore, it is important to design and offer to the user each of the following items. "The clamping method of the mold and the work-piece and the UV irradiation methods, including the direction of the irradiation at the UV imprint." "The clamping method of the mold and the work-piece, and the heating method at the thermal imprint."

Big press force is needed for thermal imprinting. Therefore, it is very important to have high stiffness, and to suppress the vibration. A cast iron design reduced the distortion occurring in the main body of the machine, even with the press force enlarged. A roller guide with high stiffness was adopted for the Z-axis (which is the press axis) to drive the AC servomotor through the ball screw, with the aim of improving the straightness in addition to the motion accuracy and positioning accuracy. Extendibility and development are necessary, in addition to ease of use by the controller. Then, the AC servomotor of the Z-axis is controlled to set the required press force, the speed of the press, and the press pattern. Moreover, the controller can set two or more pressed temperatures, the temperature rise speeds, and patterns in the thermal imprint.

Figure 6.24 shows an external view of ST50, which is a batch-processing-type machine. It can be equipped with the following as an option. "A decompression chamber aimed to suppress poor pattern transcription and bubble generation." "An XY stage for step & repeat motion (it is possible to operate this in the decompression chamber)." "An ST head® to apply the surface of the mold and the surface of the work-piece accurately (this corrects the angle by pressing)." Figure 6.25 shows the above-mentioned ST head

Figure 6.24 Externals view of ST50 (batch processing type)

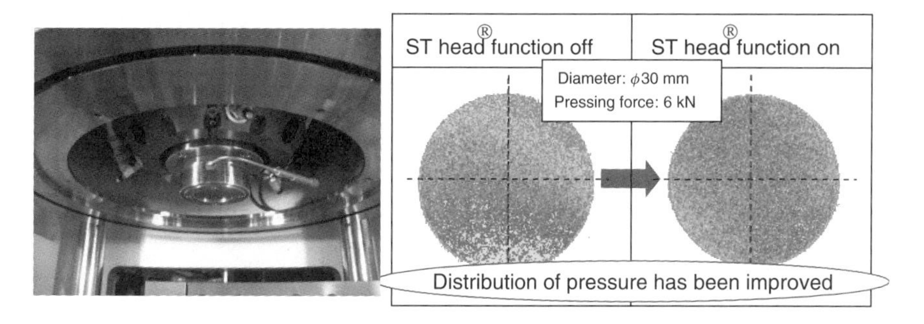

Figure 6.25 Function to adjust surface to surface: ST head®

Figure 6.26 ST table®

externals and the effect. That is, the pressure distribution when a standard mold of 30 mm diameter is held to paper, which detects the pressure. This ST head was illustrated previously in Figure 6.23(C), and it hits the "angle adjustment function" on the surface of the mold and work-piece. Moreover, the ST head of table type (ST table®) is shown in Figure 6.26 for an 8 inch diameter wafer and an A4-sized work-piece.

This ST head turns by small force, adjusting the angle, and adopts a "passive method." The important items demanded of the ST head are recorded as follows. "To locate the turn center at the center of the imprint surface." "To reduce resistance on the guide as far as possible." "When clamping, the adjusted posture is not relaxed." The most ideal methods for the above-mentioned conditions are the sphere guide, air bearing, and air clamp. However, there is a problem in the difficulty of use under the decompression condition. The ST head shown in Figure 6.25 uses a sphere guide, and ceramic ball bearing, without clamping function. Therefore, it can be used under the decompression condition. The "ST table" (sphere guide, air bearing, and air clamp) shown in Figure 6.26 can adjust the angle using very small force.

"Improvement of production time of cycle," "Correspondence to large area," "Longevity of the material for demolding," etc. are enumerated as device development keywords concerning the thermal imprint machine, and the following notes exist about each item.

Improvement of Production Time of Cycle

Given the thermal imprinting process shown in Figure 6.22(a), the heating cycle is indispensable.

"How soon is the temperature raised to the molding temperature?"
"How soon is the temperature lowered to the demolding temperature?"

Correspondence to Large Area

A very high pressure, at the tens of megapascals level, is needed for thermal imprinting. Therefore, when the area grows, a very high press force of tens to hundreds of kilonewtons is needed.

"How do we generate such a high press force?"
"Additionally, when a high press force is applied, can the accuracy of the machine be maintained?"

Longevity of the Material for Demolding

In particular, the longevity of the material for demolding is reduced when the molding temperature reaches 200 °C or more in the thermal imprinting process.

Figure 6.27 shows a high-speed heating/cooling unit with the development aim of "Improvement in production time of the cycle." In the specification, the mold and work-piece have a diameter of 4 inches, and the maximum press force is 50 kN. This unit is being developed with a heating rate target of 5 °C/s, a cooling rate target of 1.7 °C/s, and a temperature distribution in the φ 100 mm area of ± 2 °C.

6.3.2.3 Result of Thermal Imprint

An example of a thermal imprint to PMMA is shown in Figure 6.28 using an electroformed Ni mold. The specifications of the pattern are 40×40 mm^2 area, with 300 nm diameter of the hole and pitch, 200 nm in height. Such a

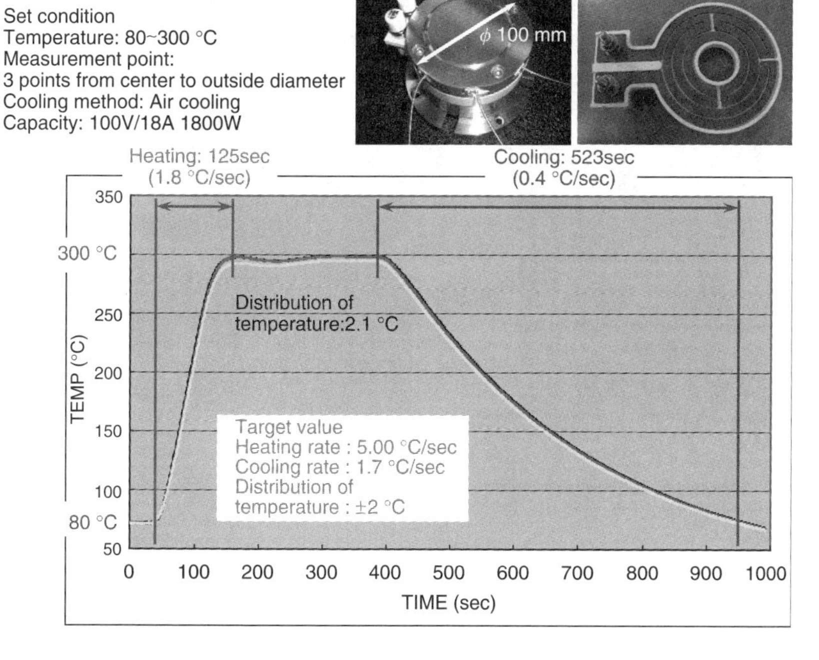

Figure 6.27 High-speed heating/cooling unit

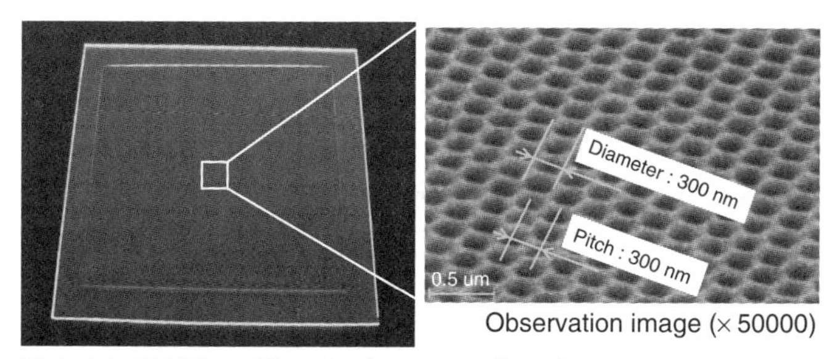

Figure 6.28 Thermal imprint sample: moth-eye structure for anti-reflection

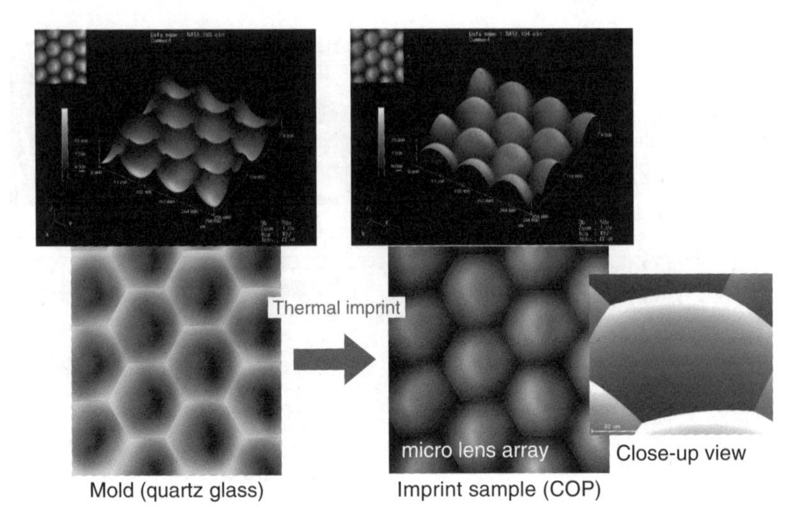

Figure 6.29 Thermal imprint sample: micro lens array structure

sample allows light to penetrate efficiently, and is used for the display, etc. Figure 6.29 shows a micro lens array of pitch about 100 μm made by the thermal imprint method. The thermal imprint was applied to cyclo-olefin polymer (COP) resin using a concave-form mold made of quartz glass.

Figure 6.30 shows another example of a thermal imprint in the passage to the substrate of a large-sized PMMA. A close-up view shows the tens or hundreds of micrometer groove and hollow sizes that can be formed even by the injection molding method. However, to form a sharp edge to the corner part when the substrate is very thin and near the film, the thermal imprinting method becomes effective. Figure 6.30(b) represents a case to use the thermal imprint chiefly for the latter reason, and an excellent sharp edge was obtained. In this case, an ST200 with press force of 200 kN was used for the thermal imprinting machine shown in Figure 6.30(a).

6.3.3 Summary

There are many hurdles to be overcome besides the choice of machine when designing a practical nanoimprint method. For the mold, accuracy, area (size), material, cost, etc. are the important factors. For the resin, easy spreading, easy molding, etch proofing, and cost. For the demolding material, demolding by weak force, easy spreading, ease of handling, and durability.

However, experimental examination of a device which uses a nanoimprint process is divided from the viewpoint of "Only the method with good

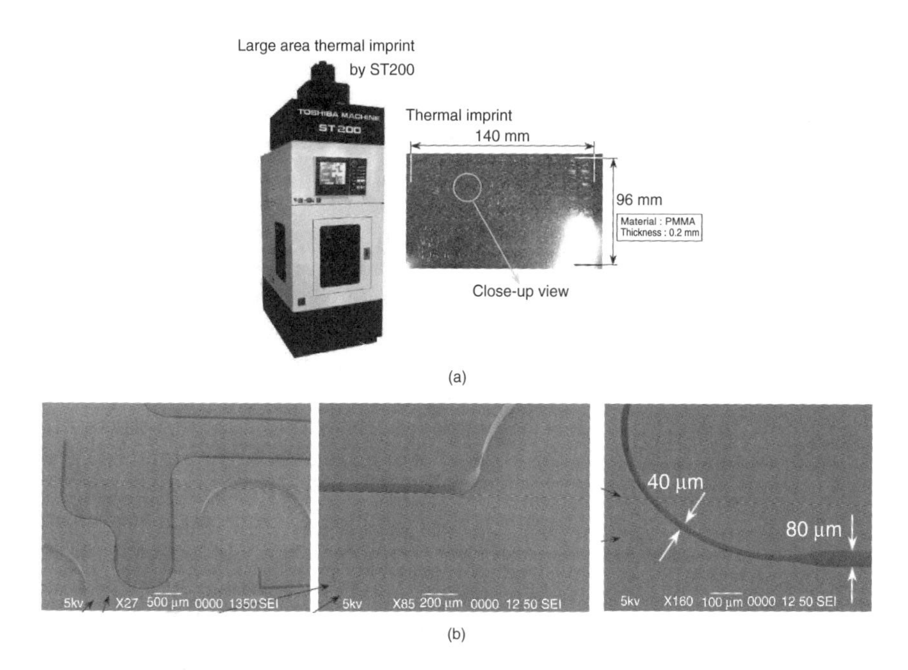

Figure 6.30 (a,b) Thermal imprint sample: bio category, passage on PMMA substrate

accuracy should be used" to "Experimental examination is done considering the cost of the device produced, the durability of the mold, and the durability of the demolding material." Therefore, we think the day is near when the nanoimprint method and machine will be applied to the production line.

Needless to say, a finished degree of improvement of the nanoimprint machine is necessary to achieve device production. Additionally, it is important to improve the technology, mold, resin material, and process design and inspection – requiring a series of manufacturing processes with similar finished result. Finally, offering these technologies in total becomes indispensable.

6.4 Micro/Nano Melt Transcription Molding Process

6.4.1 Outline of the Melt Transcription Molding Process

The micro/nano melt transcription molding process (MTM) developed by the Japan Steel Works, Ltd has features suitable for fabricating thermoplastic products having precise microstructure (ranging from tens of nanometers to hundreds of micrometers) with high aspect ratio, thin wall, large area, minimal birefringence, and low residual stress [32]. An outline of MTM is shown schematically in Figure 6.31. MTM consists of a few characteristic stages compared with thermal cycle NIL.

(1) Polymer pellets are melted by the plasticating equipment, including a rotating screw. This enables one to obtain a uniform molten polymer quickly, and reduce the time required for fabricating a molded product. Furthermore, the plasticating equipment enables the use of many kinds of thermoplastic polymer pellets widely circulated in the market.
(2) In the coating stage, a molten polymer is coated on the surface of a stamp with approximately the same length, width, and thickness of the final product. The temperature of the stamp is kept higher than the glass transition temperature (T_g) of the polymer during the coating stage. This enables the coated polymer to avoid significant temperature drop and solidification, and assures uniform coating over the whole surface of the stamp under the low and uniform coating pressure which is independent of the size of coating area, as shown in Figure 6.32. In addition, the molten polymer is consecutively coated from one edge to another on the stamp surface, so almost all the air on the stamp surface is compulsorily pushed out. This means that the efficient transcription of the microstructure can be achieved without vacuum pumping.
(3) Immediately after the coating stage, the coated polymer on the stamp is compressed as shown in Figure 6.33. Since the polymer is still flowable at the beginning of this stage, the transcription of the microstructure of the stamp surface to the polymer is enhanced even under the relatively low compression pressure. In addition, the polymer keeps its shape (hardly flowing in the cavity) during compression, because the shape of the coated polymer has almost the same dimensions as the final product. Therefore, residual birefringence, as a result of polymer molecule deformation, hardly remains in the final product.

6.4.2 High Transcriptability

To verify the molding ability of thermoplastic products with precise microstructure, thin wall, and large area, molded products were fabricated by MTM. The material used was a film-grade polymethylmethacrylate (PMMA) with high viscosity (Parapet EH, $T_g = 100\,°C$, melt mass flow rate (MFR) = 1.3 g/10 min (ISO1133, 230 °C, 37.3 N), Kuraray Co., Ltd). The temperature of molten PMMA was 280 °C. The molten PMMA was coated

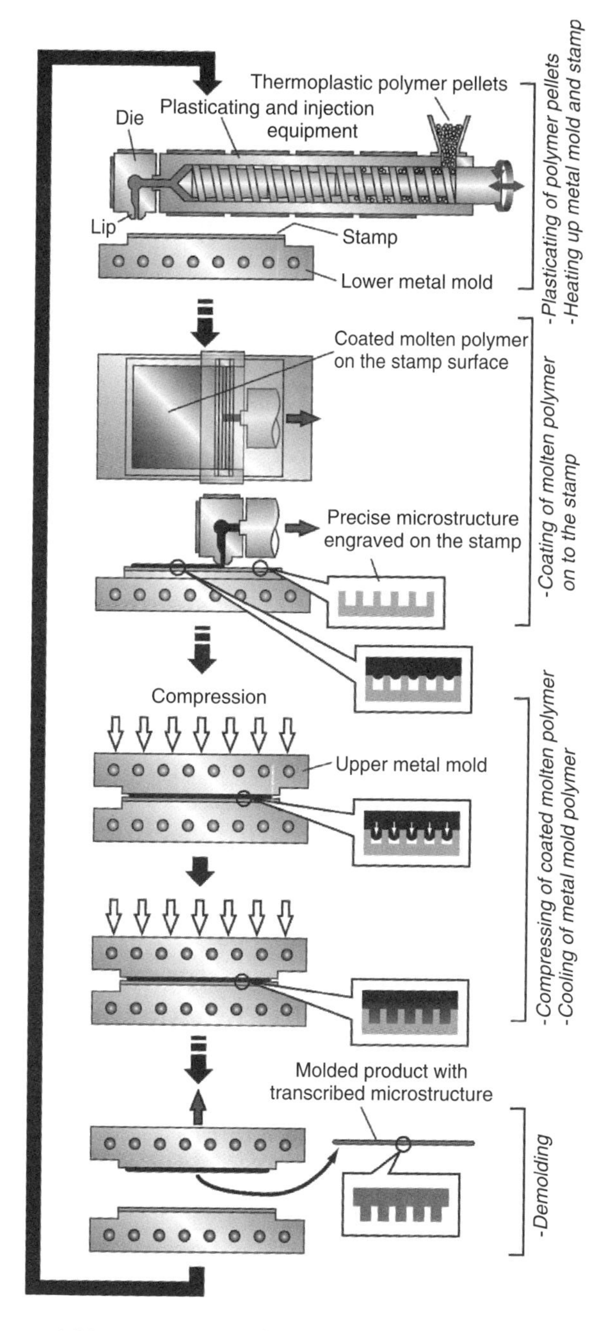

Figure 6.31 Outline of melt transcription molding process

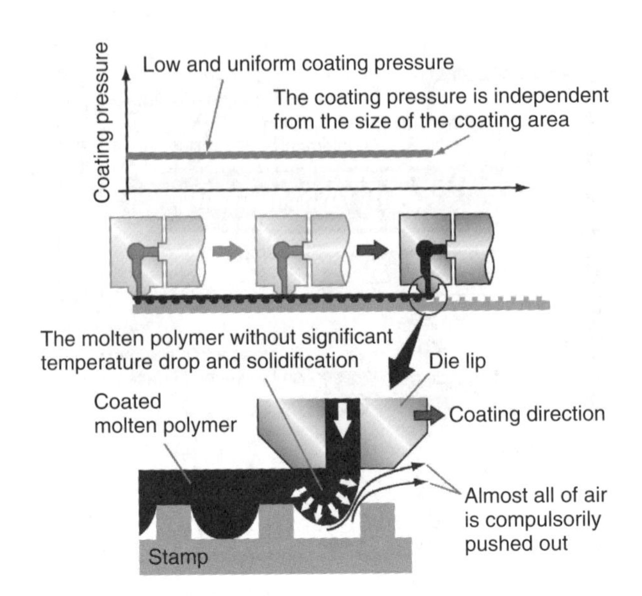

Figure 6.32 Advantages of MTM during the coating stage

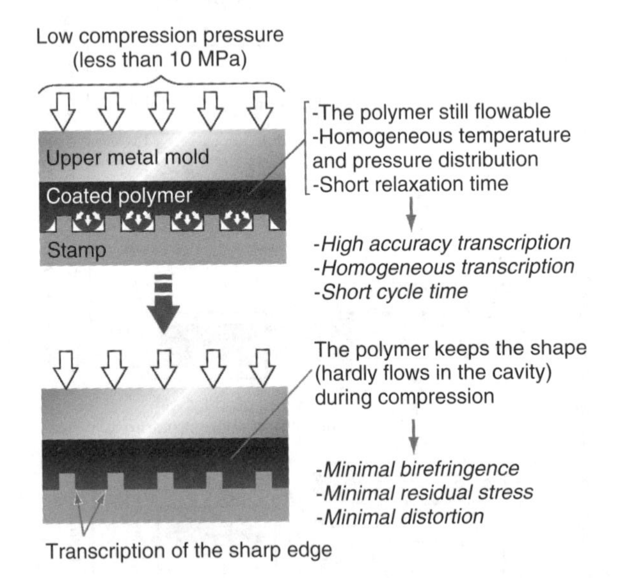

Figure 6.33 Advantages of MTM during the compression stage

over the stamp heated at 160 °C. Immediately after coating, PMMA was compressed at 11 MPa for 15 s. After that, the metal mold and PMMA were cooled below T_g. Finally, the metal mold was opened and the molded product was released from the metal mold. The cycle time was 105 s. Figure 6.34(a) shows a schematic of the PMMA product, 50 mm in width, 100 mm in length, and 200 μm in thickness. About 2.5 million minute pillars with 10 μm diameter, 30 μm height, and 20 μm pitch were successfully transcribed over the whole surface of two pattern areas, 36 mm in diameter. Figure 6.34(b)–(d) shows SEM images of transcribed pillars and logotypes. The minute pillars were uniformly transcribed on the whole area. Furthermore, the transcribed pillars and logotypes showed sharp corners

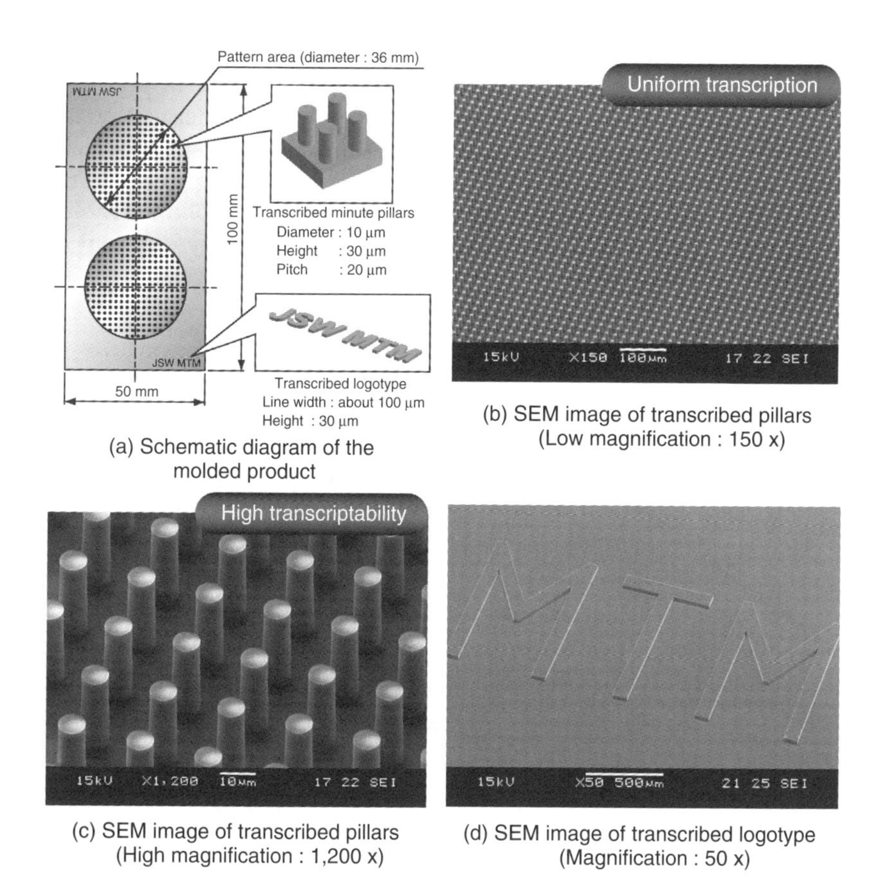

(a) Schematic diagram of the molded product

(b) SEM image of transcribed pillars (Low magnification : 150 x)

(c) SEM image of transcribed pillars (High magnification : 1,200 x)

(d) SEM image of transcribed logotype (Magnification : 50 x)

Figure 6.34 PMMA molded product with transcribed minute pillars and logotype

without any damage or bending. These results show that the MTM has a high capability to transcribe the microstructure, even if a polymer with higher viscosity is used.

6.4.3 Excellent Optical Properties

To verify the molding ability of a thermoplastic product having excellent optical properties, molded products were fabricated by MTM. The material used was a high fluidity grade polycarbonate (PC) (Tarflon IV1900R, $T_g = 145\,°C$, melt volume flow rate (MVR) = 19 $cm^3/10$ min (ISO1133, 300 °C, 11.8 N), Idemitsu Kosan Co., Ltd). Figure 6.35 shows polarized light transmission images through the PC products, 90 mm^2 square, 1.6 mm in thickness. In case of injection molded product, the optical distortion remained especially near the gate. Figure 6.36 shows a comparison of the retardation between the MTM product and the injection molded product. The MTM product has excellent optical properties. In this process, the polymer undergoes no large-scale deformation during molding, except for the coating stage in which the polymer melt temperature is high enough for the polymer molecule relaxation. Therefore, residual retardation, as a result of the polymer molecule deformation, hardly remains in the products, which is very different from the injection molding process. This feature is effective especially when the products are used for optical applications.

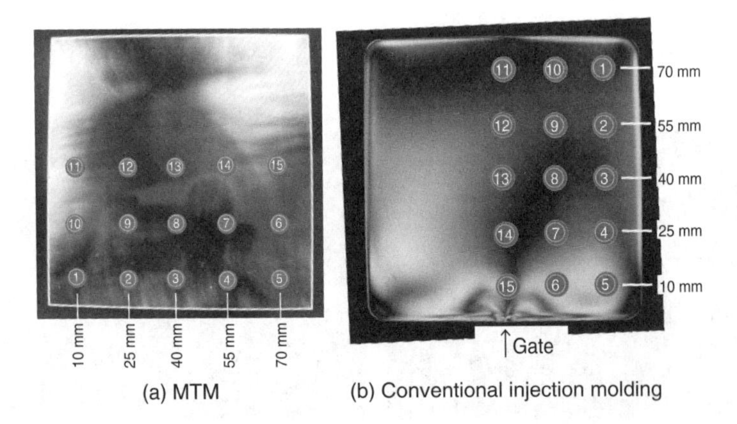

Figure 6.35 Polarized light transmission images of PC products

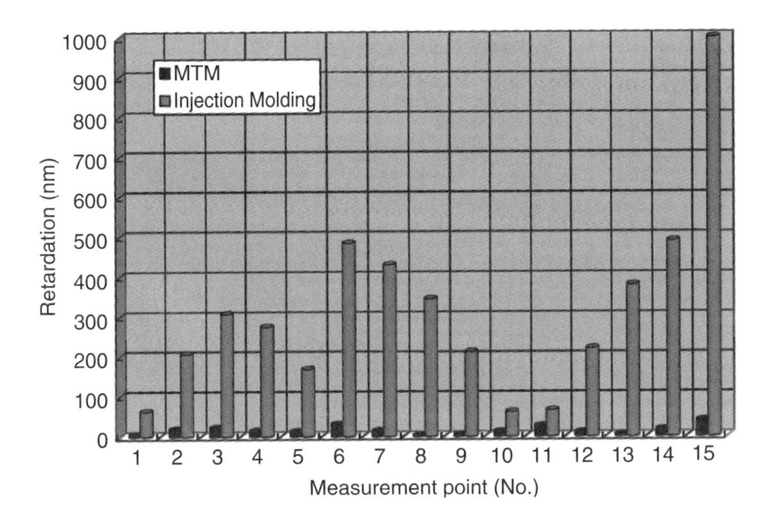

Figure 6.36 Comparison of retardations between the MTM product and the injection molded product (material used: PC)

6.4.4 Melt Transcription Molding System "MTM100-15"

Figure 6.37 shows the melt transcription molding system "MTM100-15" manufactured by the Japan Steel Works. Table 6.1 summarizes the experimental results obtained using MTM100-15. A wide range of thermoplastic polymers can be used, not only amorphous polymers (PMMA, PC, COP, COC, etc.) but also crystalline polymers (PEN, PET, etc.). The MTM was used to produce the tabular molded products up to L128 mm × W100 mm, and wall thickness from 50 μm to 1.5 mm. The microstructure fabricated on the surface of the molded products has shape ranging from tens of nanometers to hundreds of micrometers. Furthermore, a high aspect ratio of more than 10 could be obtained.

The relationships between the aspect ratio and the dimension of the transcribed microstructure fabricated by MTM, thermal cycle NIL, and conventional injection molding are shown in Figure 6.38. Black circles represent the dimensions of the typical micro-/nanostructures that have been obtained using MTM100-15. These products involve biomedical chips, optical memory for data storage, and optical nano devices such as low-reflective moth-eye structures.

Figure 6.37 MTM100-15

Table 6.1 Summary of the experimental results obtained using MTM110-15

Item	Performance
Molding material	Thermoplastic polymer (Pellets) PMMA, PC, COP, COC, etc. PEN, PET, etc.
Size of final product	48 × 48 mm 100 × 100 mm 128 × 100 mm
Thickness of final product	50 μm ~ 1.5 mm
Aspect ratio of microstructure (Height / width or diameter)	~10

6.5 Future Trends

Recently, the amount of research and development on element nanoimprint technologies, including molding, demolding, and removal of residual layer,

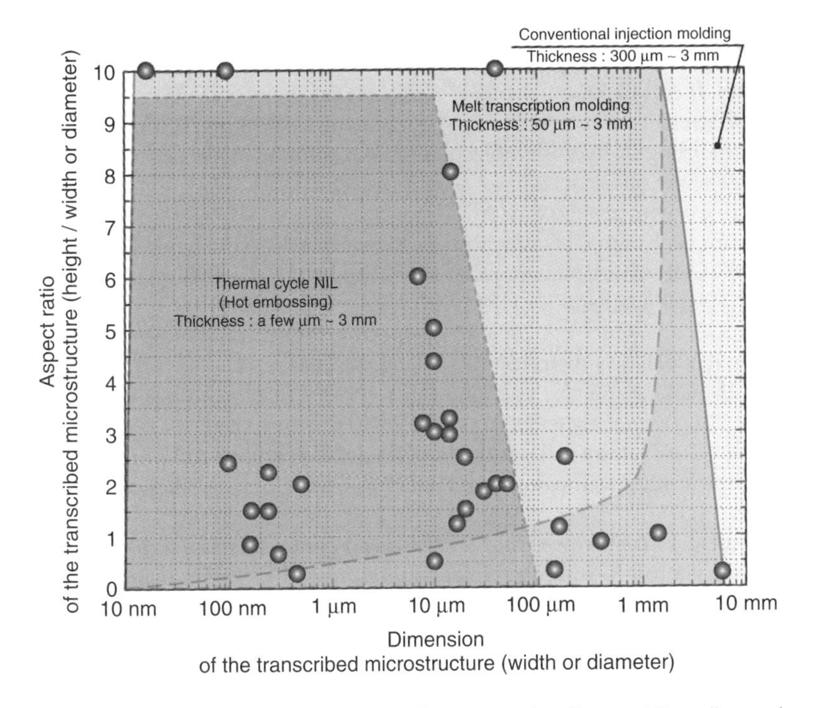

Figure 6.38 The relationships between the aspect ratio and the dimension of the transcribed microstructure

has been increasing. Resin materials for nanoimprinting have also been developed. The imprint resin requires several functionalities, including high resolution, good demolding, and high durability against electrodeposition and dry etching. In contrast, the imprint process has been applied for many electronic, optical, and chemical/biochemical devices. If the imprint resin could be used directly as part of the actual devices, then a simple and high-throughput fabrication process would be realized. Thus, the imprint resin also requires other functionalities for actual devices, such as high/low *k*, electrical conductivity, and biocompatibility. This will necessitate joint research between material processes and device developers.

References

[1] Austin, M.D., Ge, H., Wu, W., Li, M., Yu, Z., Wasserman, D. *et al.* 2004. Fabrication of 5 nm linewidth and 14 nm pitch features by nanoimprint lithography. *Appl. Phys. Lett.* **84**: 5299.

[2] Li, W.D., Wu, W., and Williams, R.S. 2012. Combined helium ion beam and nanoimprint lithography attains 4 nm halfpitch dense patterns. *J. Vac. Sci. Technol. B* **30**: 06F304.
[3] Hua, F., Sun, Y., Gaur, A., Meitl, M.A., Bilhaut, L., Rotkina, L. *et al.* 2004. Polymer imprint lithography with molecular-scale resolution. *Nano Lett.* **4**(12): 2467.
[4] http://www.obducat.com.
[5] http://www.evgroup.com/en.
[6] Nomura, S., Kojima, H., Ohyabu, Y., Kuwabara, K., Miyauchi, A., and Uemura, T. 2005. Cell culture on nanopillar sheet: Study of HeLa cells on nanopillar sheet. *Jpn. J. Appl. Phys.* **44**(37): L1184.
[7] Ahn, S.H. and Guo, L.J. 2008. High-speed roll-to-roll nanoimprint lithography on flexible plastic substrates. *Adv. Mater.* **20**: 2044.
[8] Tan, H., Gilbertson, A., and Chou, S.Y. 1998. Roller nanoimprint lithography. *J. Vac. Sci. Technol. B* **16**: 3926.
[9] Mäkelä, T., Haatainen, T., and Ahopelto, J. 2011. Roll-to-roll printed gratings in cellulose acetate web using novel nanoimprinting device. *Microelectron. Eng.* **88**: 2045.
[10] Taniguchi, J., Yoshikawa, H., Tazaki, G., and Zento, T. 2012. High-density pattern transfer via roll-to-roll ultraviolet nanoimprint lithography using replica mold. *J. Vac. Sci. Technol. B* **30**: 06FB07.
[11] Khang, D.Y. and Lee, H.H. 2004. Sub-100 nm patterning with an amorphous fluoropolymer mold. *Langmuir* **20**: 2445.
[12] Guo, L.J. 2007. Nanoimprint lithography: Methods and material requirements. *Adv. Mater.* **19**: 495.
[13] Taniguchi, J. and Aratani, M. 2009. Fabrication of a seamless roll mold by direct writing with an electron beam on a rotating cylindrical substrate. *J. Vac. Sci. Technol. B* **27**: 2841.
[14] Taniguchi, J., Unno, N., and Maruyama, H. 2011. Large-diameter roll mold fabrication method using a small-diameter quartz roll mold and UV nanoimprint lithography. *J. Vac. Sci. Technol. B* **29**: 06FC08.
[15] Kataza, S., Ishibashi, K., Kokubo, M., Goto, H., Mizuno, J., and Shoji, S. 2009. Seamless pattern fabrication of large-area nanostructures using ultraviolet nanoimprint lithography. *Jpn. J. Appl. Phys.* **48**(6): 06FH21.
[16] Hiroshima, H., Inoue, S., Kasahara, N., Taniguchi, J., Miyamoto, I., and Komuro, M. 2002. Uniformity in patterns imprinted using photo-curable liquid polymer. *Jpn. J. Appl. Phys.* **41**(6B): 4173–4177.
[17] Choi, J., Nordquist, K., Cherala, A., Casoose, L., Gehoski, K., Dauksher, W.J. *et al.* 2005. Distortion and overlay performance of UV step and repeat imprint lithography. *Microelectron. Eng.* **78/79**: 633–640.
[18] Lee, E., Jeong, J., Sim, Y., Kim, K., Choi, D., and Choi, J. 2006. High-throughput step-and-repeat UV-nanoimprint lithography. *Curr. Appl. Phys.* **6**: e92–e98.
[19] Nagai, N., Ono, H., Sakuma, K., Saito, M., Mizuno, J., and Shoji, S. 2009. Copper multilayer interconnection using ultraviolet nanoimprint lithography with a double-deck mold and electroplating. *Jpn. J. Appl. Phys.* **48**(11): 115001.

[20] Chao, B., Palmieri, F., Jen, W.L., McMichael, D.H., Willson, C.G., Owens, J. *et al.* 2008. Dual damascene BEOL processing using multilevel step and flash imprint lithography. Proc. SPIE, Vol. 6921.
[21] Ono, H., Ono, Y., Kasahara, K., Mizuno, J., and Shoji, S. 2008. Fabrication of high-intensity light-emitting diodes using nanostructures by ultraviolet nanoimprint lithography and electrodeposition. *Jpn. J. Appl. Phys.* **47**(2): 933–935.
[22] Bergh, A.A., Hill, M., Saul, R.H., and Plains, S. 2002. United States patent US 3739217.
[23] Fujimoto, A. and Asakawa, K. 2007. Nano-structured surface fabrication for higher luminescent LED by self-assembled block copolymer lithography. *J. Photopolym. Sci. Technol.* **20**(4): 499–503.
[24] Oder, T.N., Kim, K.H., Lin, J.Y., and Jiang, H.X. 2004. III-nitride blue and ultraviolet photonic crystal light emitting diodes. *Appl. Phys. Lett.* **84**(4): 466–468.
[25] Fujii, T., Gao, Y., Sharma, R., Hu, E.L., DenBaars, S.P., and Nakamura, S. 2004. Increase in the extraction efficiency of GaN-based light-emitting diodes via surface roughening. *Appl. Phys. Lett.* **84**(6): 855–857.
[26] Wierer, J.J., Krames, M.R., Epler, J.E., Gardner, N.F., and Craford, M.G. 2004. InGaN/GaN quantum-well heterostructure light-emitting diodes employing photonic crystal structures. *Appl. Phys. Lett.* **84**(19): 3885–3887.
[27] Orita, K., Tamura, S., Takizawa, T., Ueda, T., Yuri, M., Takigawa, S. *et al.* 2004. High-extraction-efficiency blue light-emitting diode using extended-pitch photonic crystal. *Jpn. J. Appl. Phys.* **43**(8B): 5809–5813.
[28] Huang, H.W., Kao, C.C., Chu, J.T., Kuo, H.C., Wang, S.C., and Yu, C.C. 2005. Improvement of InGaN–GaN light-emitting diode performance with a nano-roughened p-GaN surface. *IEEE Photon. Technol. Lett.* **17**(5): 983–985.
[29] Guo, L.J. 2004. Recent progress in nanoimprint technology and its applications. *J. Phys. D: Appl. Phys.* **37**: R123–R141.
[30] Chou, S.Y. *et al.* 1996. *J. Vac. Sci. Technol. B* **14**(6): 4129.
[31] Goto, H. *et al.* 2007. Micro patterning using UV-nanoimprint process. *J. Photopolym. Sci. Technol.* **20**(4): 559–562.
[32] Ito, H., Satoh, I., Saito, T., and Yakemoto, K. 2007. *Int. Polym. Process.* **XX**(2): 155.

Index

Nanoimprint Technology: Nanotransfer for Thermoplastic and Photocurable Polymers, First Edition.
Edited by Jun Taniguchi, Hiroshi Ito, Jun Mizuno, and Takushi Saito.